D0948343

CAREERS IN CABLE TV

A complete guide to
getting a job – from
receptionist to producer
in America's fastest-
growing entertainment
industry

JON S. DENNY

BH EH 588 $7.95

CAREERS IN CABLE TV

JON S. DENNY

BARNES & NOBLE BOOKS
A DIVISION OF HARPER & ROW, PUBLISHERS
New York, Cambridge, Philadelphia, San Francisco
London, Mexico City, São Paulo, Sydney

FIRST EDITION

Designer: C. Linda Dingler

Library of Congress Cataloging in Publication Data

Denny, Jon S.
  Careers in cable TV.

  (Everyday handbook; EH/588)
  Includes index.
  1. Cable television—Vocational guidance. I. Title.
HE8700.7.C6D46   1983   384.55'56'02373   82-48660
ISBN 0-06-463588-0 (pbk.)

---

83 84 85 86 87 10 9 8 7 6 5 4 3 2 1

CONTENTS

ACKNOWLEDGMENTS

This book—and my life, for that matter—could not have been possible without the attraction that my parents have for each other. Albert and Annette provided much support and sentience, but they still won't subscribe to cable TV. A major nod to my editors Larry Ashmead and Craig Nelson, who know how to shape and inspire, even though they gave more money to Saul Bellow. A hello to Stephanie Anderson, Edward Bott, and Debbie Trontz—crack researchers all. A kudo to the guy who fixes my typewriter. A kiss to Pamella Greene, my LG, who now knows more about cable than she ever wanted to. A handshake to Ted Turner, who knows the meaning of the words "courage" and "moustache wax." A rub on the head to my dog Buddha, who thinks Rin-Tin-Tin should have his own network. Applause to all the gray-haired gents who thought cable TV was a damn good way to get reception. And many thanks to the countless number of cablephiles that I interviewed for the book. Now, get back to work!

INTRODUCTION

The best thing that can be said about cable television is that it has no tradition. That's a cheery piece of reality for people starting to focus on a career. In cable TV, there is no tradition for hiring, no hierarchy, no "old boy network." There is an unswerving demand for people who know a bit more than the difference between cable TV and TV dinners.

Because it is a relatively new industry, with few experienced hands, cable is a comparatively open field for women, the unskilled, and the young at heart and in age. The average age of the cable employee is under thirty-five. Entry-level positions are available in almost every facet of the industry, with most cable companies expressing a "crying need" for technical people and a serious interest in people with degrees in business, English, or communications. There is a profound need for people who have the verve to service and sell. Fact: According to the National Cable Television Association, 1,000 new people get hired to work in cable every month.

What began as a means of transporting clear television signals to rural areas is now a burgeoning industry drawing 250,000 more subscribers every month. And as more subscribers hook up and more cities grant cable franchises, career opportunities will continue to increase. Statistics reveal that cable TV reaches 40 percent of all television homes in the United States. By the end of the 1980s, penetration could be at least 80 percent!

Another interesting number: 46 percent of those homes currently able to sign up for cable service choose *not* to subscribe. That underscores the pressing need in the consumer-driven cable industry for people who have a grasp of advertising, promotion, and marketing. Cable television must learn how to deal with its public.

Television in America is a legal narcotic and cable TV has enhanced the fix. There are over forty cable satellite programmers, and names and acronyms such as HBO, Showtime, ESPN, and MTV mean that there are plenty of programs and job opportunities. There are close to 6,000 cable systems in more than 13,000 communities, which means that there could be a golden opportunity not far from your own backyard.

"How do I get into cable TV?" That's a question that has been put to me a dozen or more times, by warm bodies and bright minds interested in slicing off a piece of the video pie. It is a question that has been posed hundreds and thousands of times to cable professionals who have carved out a place on the ground floor and are now taking the career escalator up and up. "How do I get in?" "What's the best way to go?" "Where are the real jobs in cable TV?"

In the beginning of a career in cable television, the hardest thing to see is the industry as a whole. It is difficult to channel your interests into a job when you don't know where to turn. It is almost impossible to get an overview—a prismatic *feel* for the industry and the opportunities that exist.

It is the purpose of this book to provide the overview and feel. It is a book which will help you lay out the puzzle and figure out where you fit in. It is a book intended to help you become a working member of the cable community.

I would have appreciated a book of this sort when I first started out in cable TV. I recall the days and daze of searching for an edge in, armed only with a few good ideas, a great deal of energy, an attaché case, and a belief that cable was the place to be if all you had to sell was yourself. I remember traveling a thousand miles with my partner to a place called Sumter, South Carolina, on the promise of production money for an early concept for cable TV. We ended up paying for the lunch of the potential investor, and the frustrating part was that the man owned the restaurant! Needless to say, we didn't get a dime.

I recall a bunch of wasted hours, but I learned from them, too. I remember thinking, after having created, sold, and produced my first special for a national pay-cable network at age twenty-four, how quixotic I had been in the beginning. When it comes down to it, there is absolutely nothing quixotic about making a deal or finding a job in cable TV. But that doesn't mean that you can't live out your dreams.

Disraeli said it best: "The secret of success in life is for a man to be ready for his opportunity when it comes." This book has been written to help people prepare for the opportunities in cable TV. If you are just starting out—or if you are already on your way— it will provide you with a vantage point and, it is hoped, a good deal of insight. It will tell you where the jobs exist and what it takes to do them right.

Since cable television is so new, there are chapters which will pinpoint exactly what it is, where it's been and in what direction it's going. That's all part of the much-needed overview. The many career profiles throughout the book will allow you to see the step-by-step progress involved in building a successful career in cable. The profiles will also help you take a practical, knowledgeable approach to the cable job market. The section on proceeding—and succeeding—with that first giant step sifts out tips from the cable professionals who do the hiring. The section on cable access tells you how to become a cable superstar if you want to, and where to learn production skills if you desire to make it behind the scenes. The appendix contains the kind of invaluable contacts and information which you will want to grab onto for future reference. The glossary teaches the language of the cable pro. In between, there's a chapter called "An Education in Cable TV"—which describes various intern programs, workshops, and cable courses, a plethora of job descriptions, and the insight and advice from the experts which will get you off to a running start. The cable television industry has room with a view for people who know how to use their talents. The rapidly evolving industry has a demand for people who know where—and why—they can fit in. It is my hope that this book provides the necessary direction and handle. It is a book that has been written for the soon-to-be cablephile who wants and needs an "in."

ONE

What Is Cable TV?

If you subscribe to one of the new, state-of-the-art, very expensive cable systems, you're undoubtedly familiar with the syndrome: You talk to a salesperson, and he's not sure what the difference is between Showtime and the Movie Channel; Home Box Office suddenly dissolves into a garbled mishmash of colors and snow, and the service technician assures you that your service will be restored —just as soon as they find the problem. The fact is, much cable technology is so recent that even some of the veterans of the industry aren't yet entirely comfortable with it.

In Southern California not long ago, a young man applied for a job with a cable company as a salesman. His qualifications were excellent, his energy level was high, and the interview went well, but he didn't get the job. The reason? He kept referring to his old employer—a Southern California–based *over-the-air* subscription TV service—as "that other cable company." These days, cable operators are leery to hire new people who don't know at least the basics of cable technology and what separates it from its noncable competitors. Fortunately, learning how cable TV works doesn't take an advanced engineering degree, just an understanding of the chain of manpower and high-tech hardware that brings the incredible assortment of television signals into America's living rooms.

Cablecasting begins with programming, often bounced off a satellite for simultaneous reception across the country, and ends at your television set. But the heart and soul of any cable system is the *headend:* the collection of sophisticated equipment and trained technicians who gather incoming signals and distribute them along coaxial or fiber-optic cable to the subscribers who've paid a fee for the service.*

Before even a single inch of cable is strung, years of work go into setting up a modern cable system. The biggest hurdle is acquiring a franchise in the first place. Like the telephone company, cable operators have to apply for the right to service a specific area. Competition is fierce, and the stakes are high for the most lucrative big-city franchises, so it's not surprising that huge *multiple system operators* (MSOs), owned by such corporate giants as General Electric, Times-Mirror Corp., and Warner-Amex, have pumped large sums into the race for those territories. Just to enter a bid and follow through on it, insiders say, will cost more than $2 million —and the market is not bottoming out. These days, municipalities are likely to award a franchise to the company that promises the flashiest and most sophisticated system, and the one that plays the best politics.

Once the franchise is won, the real work of building the system begins. Needless to say, the biggest ingredient is cash—a ton of it. Millions of dollars worth of expensive receiving, transmitting, and switching equipment is needed, along with trained personnel and the physical plant to house them. Then there's the cost of running the cable itself. In 1982, the National Cable Television Association (NCTA) estimated that stringing cable along telephone poles can cost between $7,500 and $15,000 per mile. That cost can spiral to more than $100,000 a mile if underground installation is required. Finally, factor in the cost of individual cable convertors for each subscriber ($35 and up), sales forces to bring in new customers, and service technicians to handle the inevitable problems, and the price

*For definitions of these and other technical cable terms, see the glossary, p. 239.

tag can soar. Four years ago, Times-Mirror Corp. paid $400 a subscriber to purchase an existing cable system; analysts estimate that building that system from scratch today would cost a cool $1,000 per subscriber.

With the system in place and subscribers anxiously waiting for programming to be delivered, the actual process of cablecasting begins. The first step is assembling a package of programming to fill the system's channel capacity. In general, there are five types of signals for the operator to choose from:

1. *Local broadcast signals.* The most basic form of cable service has changed little since the medium's earliest days. Strong signals are picked up using a large antenna and sent along the cable.

2. *Distant signals.* Stations in outlying areas with an over-the-air signal that is too weak for normal reception can amplify their broadcasts and beam them using microwave transmitters. The highly directional transmissions are picked up at the headend using small "dish" antennas.

3. *Satellite signals.* Today, cable's biggest drawing card is the proliferation of services delivered solely by satellite. A single receiving antenna aimed skyward can pluck as many as twenty-four separate program sources off one "bird." To receive signals from other satellites, however, the operator must invest in separate receiving dishes. If the satellite signal is scrambled electronically, more equipment is required for decoding before it is passed on to subscribers.

4. *Local origination* (LO). With the right equipment, a cable system can transmit its own programming along the cable. These signals can begin as videotapes, film, or even direct from a studio at the system's headend. The most popular form of local origination is public access and educational programming. Some systems, however, have even begun their own premium movie channels using this method. Group W Cable's Z Channel in Los Angeles is the best-known example. On Long Island, cable subscribers can order a stand-alone pay service called Montage, which is not available on any other system. Tulsa Cable has Movie Five; and in San

Jose, cable subscribers can opt for the G Channel.

5. *Telephone transmissions.* Although the quality of these signals is inadequate for most video needs, simple alphanumeric displays can easily be sent over the phone. The biggest users of this technology are videotex services supplying data such as "electronic newspapers" and program listings.

Juggling these pieces into a comprehensive package of cable services is a difficult job that can be made easier with the right technology. Other factors affecting the puzzle include local regulations imposed on the franchise, cable channel capacity, and the cost of individual services.

The next task for the system operator is getting the package of services to subscribers. Designing, installing, and operating the equipment to get this job done requires years of training and experience; fortunately, understanding the process is far simpler.

At the headend, signals arrive in a variety of forms: The local NBC affiliate, for example, may transmit over Channel 4, while satellite signals usually come in as direct audiovideo impulses. At the headend, all incoming signals are attached to carrier frequencies corresponding to a station number (or letter) on the channel selector box. If the system offers a premium service like Home Box Office or Showtime, the signal may be scrambled so that only paying customers can receive it. Usually, computers are used to control this switching process so that each program winds up going out on its assigned frequency.

This barrage of electronic information leaves the headend on a trunk line covering the entire franchise area. At selected strategic points along its route, feeder lines split off from the trunk to reach individual groups of homes and apartments. The final branch in this electronic tree is the dropline: a short cable run from the feeder line into a subscriber's home. If the installation is of sufficiently high quality, the signal entering the home should have changed very little from the one leaving the headend.

The final link between the headend and the television watcher is the convertor box. Without this attachment, the typical cable subscriber can receive only those signals allotted over regular broadcast channels. The convertor allows reception of signals in the midband and superband—a group of frequencies clustered between Channels 6 and 7 and above Channel 13. In older systems, the convertor itself contains the electronic circuits that make reception of premium programming possible. More modern systems are "addressable," that is, the convertor's functions can be controlled using a computer at the headend. Addressable systems also make two-way communications like Warner Amex's Qube system possible.

Cable TV may be the best technology to deliver a wide range of television signals, but it certainly is not the only alternative to conventional through-the-air broadcasting. What makes the new video technology even more confusing is that some of these alphabet-soup systems have the potential to compete with some cable systems and to augment others. Let's examine the most common noncable video delivery systems:

Subscription Television (STV) Technically, any service that charges a monthly free to deliver programming is STV. But the term is usually applied to the noncable forms of pay-TV: premium services that use conventional UHF broadcast channels to deliver a scrambled signal and then charge a monthly fee for the decoder. Oak Industries' ON TV, for example, broadcasts on Channel 52 in Los Angeles, and several cable systems carry the station, although ON collects the fee for its decoders. STV services don't depend on the cable; in fact, the most successful ones thrive in areas that aren't yet wired for cable. In Dallas, an aggressively marketed STV service was so successful that Warner Amex was forced to sharply accelerate the timetable for its cable system there.

Multipoint Distribution Service (MDS) MDS operators use a fixed antenna to broadcast a high-frequency microwave signal in all directions. Homes equipped with the right antenna in a line of sight with the transmitter have no trouble receiving MDS signals.

MDS is a favorite way of delivering STV services. Currently, the major disadvantages of the MDS system are its limited channel capacity (only one, although multi-channel systems may appear soon) and its susceptibility to interference from obstructions. As of 1981, MDS systems in the United States boasted a total of 600,000 subscribers.

Direct Broadcast Satellite (DBS)   This is literally an idea that has not yet gotten off the ground, although it may not be far off. DBS entails beaming programming direct from a satellite to homes equipped with small receiving dishes or antennas located on individual rooftops or window ledges. The Federal Communications Commission (FCC) recently authorized several DBS systems, which are expected to reach apartment dwellers in 1983 and single-family homes by 1986. At present, a major drawback of DBS is its three-channel maximum capacity, although some twenty-four-channel systems are currently on the drawing boards. The idea is an exciting one, but the emerging industry is unlikely to put a dent in the growth of cable until the end of the decade.

Satellite Master Antenna Television (SMATV)   This offshoot of DBS is commonly known as "cable without the cable." Some prototype systems are already in place, but SMATV is unlikely to threaten cable operators except in sparsely populated rural areas where laying cable is economically unfeasible. Channel capacity is also a major barrier to SMATV's success in the near future.

The alphabet soup of new technologies that transmit over the air rather than through wires represents a challenge for the cable industry, but experts agree that there's more than enough room for everybody. At present, the consensus is that cable will emerge as the stronger vehicle for pay-TV because of its broader multiple-channel capacity and its inherent flexibility.

As the other technologies advance, cable will also be making some leaps forward. With several lucrative big-city franchises still up for grabs and the majority of the country not yet wired for cable, future systems are likely to take advantage of the huge channel capacity of fiber-optic cable and steady improvements in interactive cable systems to enhance their advantage over the competition. But it's helpful to note that, despite these state-of-the-art predictions, most American cable customers will continue to be served by simple twelve- and twenty-channel systems for a long time to come, with considerably less advantage over the competition than newer, sixty-channel operations.

If you're serious about a career in cable, you can take solace in two inescapable facts. First, there will be a continuing need for manpower in the industry. As long as there's cable to lay, equipment to run, and services to sell, there will be jobs in the industry. And second, the successful operators of the 1980s will be those who are willing to hire the personnel and spend the money required to promote their systems. As Tony Cox, executive vice president of Home Box Office, admits: "Cable has got to go out and sell itself to customers as much as it sells itself to cities in order to win franchises. These so-called competitors are cream-skimmers. . . . Customer loyalty to STV or the others does not compare to cable loyalty—but we still have to work at it."

TWO

The History of Cable Television

The rise of the cable television industry is a story that would make Horatio Alger beam with pride. From its humble beginnings in rural, small-town America, the cable business has sprouted into one of the largest and most influential communication forces in North America. And cable TV—the fastest-growing frontier since bottled water—has still reached only a fraction of its potential audience and importance.

In Mel Brooks's film *The Producers,* there is a scene in which the garage-sale Nazi learns that his musical *Springtime for Hitler* is about to be produced at last. "Tonight, Broadvay," he boasts; "tomorrow, da VORLD!" Cable TV has made it to Broadvay, and da vorld isn't far behind. Today, short of owning your own, custom-made earth station or possibly the Pentagon (they have a lot of earth stations), plugging into the cable is the best way to monitor what is happening around you, and the best way to sample the growing variety of programming available from distant corners of the globe.

It is also a wonderful status symbol or plaything for those of you who would rather switch than watch.

To understand how cable television became what it is today, a bit of broadcasting history is in order. The year was 1948 and the television industry, slowed by World War II, had recovered completely and was beginning to boom. For the first time ever, four networks—NBC, CBS, ABC, and the ill-fated Dumont chain—offered prime-time programming each and every night of the week. An erstwhile vaudeville performer and ready buffoon named Milton Berle was earning the nickname "Mr. Television" for his Tuesday-night antics on NBC's "Texaco Star Theatre." There were almost 1 million black-and-white TV sets sitting in living rooms and bars across the country. Americans were getting hooked on the new medium that flickered.

Viewers in New York and Los Angeles, where seven stations were on the air, were the first to experience the diversity of television programming. Chicagoans had a choice of four stations, and a few other spots, including Philadelphia, San Francisco, and Baltimore, had access to three channels. All told, about one hundred station licenses had been granted by the Federal Communications Commission in 1948, when the Commission abruptly stopped issuing new permits so it could study the technical aspects of broadcasting.

The FCC ban, which lasted until late 1953, threatened to cripple the television industry outside the major metropolises, where the bulk of station licenses had been granted. Most rural Americans found themselves unable to receive even a single satisfactory signal. All they could do was read about Milton Berle. This problem led two men, 3,000 miles away from each other, to devise extremely similar solutions.

The first—and the man most often called "the father of cable TV"—was John Walson of Mahanoy City, Pennsylvania. The modest start of John Walson's adventure in cable television illustrates the typically American feature of the industry: small-town serviceman for Pennsylvania Power & Light becomes lucky in a high-risk business. Actually, it wasn't so much good but bad luck that gave Walson his start. In fact, if it weren't for the smell of formaldehyde, John Walson might be performing appendectomies today instead of heading the largest individually owned cable system in the world.

Walson, born John Walsonavich sixty-seven years ago in Forrest City, Pennsylvania, left the state after high school and headed for Chicago's Loyola University, with the intention of becoming a physician. But one or two sniffs of the formaldehyde used to preserve the medical school's cadavers sent him in search of another career.

After completing a course of study at a nearby electrical school, Walson returned to Pennsylvania and Mahanoy City, where he took a job as a line serviceman for Pennsylvania Power & Light. While employed by P P & L, Walson became interested in electrical appliances and obtained a General Electric franchise in 1945. He continued his employment with the power company and, along with his wife Margaret, worked in his appliance store at night and on weekends. It was there, in the back of the store at Mahanoy and Pine, that the Walsons slept. "Sometimes," said Walson later, "I couldn't sleep well because I couldn't sell my television sets."

Television sets were the new order of the day, and no one was ordering. The problem was that Mahanoy City sat in a bowl surrounded by the Appalachian Mountains some eighty-five miles northwest of Philadelphia, which prevented Walson from demonstrating the striking new product called television for local residents—and potential buyers. No video signal could reach down into the city of 10,000 people, many of them coal miners with money to burn. The problem wasn't money, even though each of the 12½-inch black-and-white sets had $400 price tags. In Mahanoy City, John had a mountain to climb.

First, he tried running an antenna from the roof of his store, which was situated in the center of town. It didn't work. Walson, an inventive man who had, as a teenager, built an automatic garage door opener for his parents, then thought up the next possible solution. He drove to nearby New Boston Mountain and mounted an antenna on top of a seventy-foot utility pole. Every night, Walson would load three television sets in the back of his truck and drive prospective customers to the mountaintop for demonstrations. It was there that John Walson found the way to eliminate the snow and flutters that were turning his TV screens into a blur of blinking lights. The logistics remained a problem: "It was embarrassing to take a lady up to the mountain at night to demonstrate television," he said. "Also, some people were catching colds. I knew that if I was going to sell sets, I'd have to bring the signal into town."

On the south end of Mahanoy City, four or five blocks from the appliance store that doubled as his home, Walson had bought a warehouse that doubled as a repair shop. There, he kept his inventory of refrigerators and ovens, and the television sets that no one was buying. Right outside the rear exit of the warehouse, there was a steep hill. It was 1947; Walson picked a good-looking spot and hammered in a pole. He then purchased some heavy-duty twin-lead army surplus wire from a merchandising store in Philadelphia. This wire was run on trees from the pole that he had jammed on the high side of the mountain, right into the warehouse on the south side of town. Walson had not forgotten to connect an antenna from the top of the pole, which enabled him to scoop up the TV signals from Philadelphia—all three of them.

He still had to get the signals from his warehouse to the appliance store, where he intended to sell his sets. Watson made a left turn at the corner, went two blocks up, and two blocks down, stringing his wire on Pennsylvania Power & Light poles along the way. As the lines were strung from mountain pole to warehouse to appliance shop, Walson hooked up several homes en route. Dr. Aaron Liachowitz, a local optometrist, Frank Boyle, George Barlow, and Robert B. Gray all agreed to become cable television's first subscribers. In addition to the antenna, Walson also installed amplifiers called electro voice boosters, and positioned them every five hundred feet, to improve reception. Finally, he reached his store, placed three TV sets in the window, turned on the three different channels, and then turned down the sound of all but one channel. "I wanted three different pictures in the window," he said, "but not three different sounds."

A hundred or so curious people gathered around the storefront window, some of them sitting on lawn chairs, ready to experience television—and a clear television picture—for the first time. "The first hour was exhilarating," said Walson. "People were camping out, pointing, and watching with their eyes, ears, and mouths open." Channels 3, 6, and 10 from Philadelphia were a cause for celebration, and soon the people got up out of their chairs and walked in—asking how much and how does it work. It took Walson one afternoon to put in a half mile of wire, and "more than one," he said, "to explain how it all happened."

Nevertheless, seven of the locals became his first customers, agreeing to buy sets hooked up to his cable connection. Walson charged them $100 for the installation, which included the first year's service fee. Thereafter, the monthly service rental was $2. By June of 1948, 727 subscribers were paying $2 a month to watch television. The next year, Walson's Service Electric Company upgraded the system by replacing the home-type flat ribbon wire with more effective coaxial cable, and increased the number of paying customers to 1,500. Then, in 1952, Walson was the first person to use microwave technology to relay TV signals. Channels 5 and 11 from far away New York became available in Mahanoy City.

Shortly thereafter, Walson dropped selling refrigerators and ovens, and concentrated on cable television. It was a wise move. In 1972 he helped Home Box Office launch the first successful premium cable service in America, and by 1982 Walson had over 200,000 subscribers in two states and was recently cited by Congress for founding cable television.

Walson has faced more than his share of obstacles. The Pennsylvania Power & Light Company threatened to "throw him off" their utility poles in 1950. Robert B. Gray, one of the early subscribers, was also the local manager of P P & L, and had to admit, when questioned, that he had given John Walson "verbal permission" to attach antennas to their poles. Power & Light executives continued to talk of Walson as an unauthorized tenant, but were forced to come up with a pole attachment agreement on October 2, 1950. With the good sense of a smart businessman, Walson had wired the homes of a handful of local politicians, who didn't want anyone—including P P & L—toying with their cable service.

Walson, now sixty-seven, has also had a couple of setbacks with a recent idea he calls "smell-a-vision." "It's a system in which attachments to the television set will emit the aromas of products endorsed on the screen." So far, Walson hasn't been able to encourage much interest in the concept.

He hadn't done so badly on the olfactory front, however. If not for the smell of formaldehyde, remember, John Walson would never have come up with cable TV.

John Walson of Mahanoy City was the first, but he was quickly followed—whether they knew it or not—by a string of hardy entrepreneurs aiming to improve television reception so they could sell the sets. Across the country, in Astoria, Oregon, a man named L. E. Parsons (who usually answered to the name of Ed) owned a local radio station and—no surprise—an appliance store, and had been bitten by the TV bug at a Chicago convention in 1947. When he considered television for Astoria, he came face to face with several offending mountains that blocked reception from Seattle, about 150 miles away.

After much tinkering, which left him with radio but still no TV, Parsons discovered that the rooftop of the eight-story Astor Hotel downtown offered the cleanest shot at a television signal. With the permission of the hotel owners. Parsons installed an antenna at the top of the Astor and ran a wire down the side of the building and along a few city blocks to his home.

A few tinkers later, Parsons tried the new-fangled invention in his own home. "Reception wasn't of a quality that would be salable today," he said, "but we received a picture and started inviting guests." It was Thanksgiving 1948, and the occasion was the first broadcast of KSRC, Seattle's Channel 5. Soon, there were so many guests that the Parsons' small living room overflowed with people. Ed Parsons's solution was to run separate lines to the hotel lobby and to a music store across the street.

Like Walson, Parsons began installing cable hook-ups to his neighbors for an average charge of $100, but his system may have been too sophisticated for its time. Customers owned the antennas and lines cooperatively, and the signals were fed from a melange of separate amplifiers and antennas scattered about town. By May 1950 the total number of television homes in Astoria, Oregon, was only about two hundred, and early the following year Parsons had to raise his rates: to $125 for an installation and $3 a month for a service charge. At the end of 1952, Parsons's business was in receivership. It was purchased by a group of local businessmen.

Parsons was more successful as a consultant and manufacturer, selling antennas, cable wire, and a bit of know-how to entrepreneurs from around the country. Night after night, a familiar scene in towns all over the nation materialized. Dozens of locals would gather to watch the new miracle of television.

The market was ripe, and new applications and technology were developing each year. Milton Jerrold Shapp, founder of Jerrold Electronics Corp. and the former governor of Pennsylvania, first demonstrated a master antenna system at a 1949 broadcasting convention. In 1951, a group of ten cable pioneers met at the Necco-Allen Hotel in Pottsville, Pennsylvania, and formed the National Community Television Association (NCTA). Many, including Ma Bell, the networks, and the FCC, were convinced that as soon as the TV channel allocation freeze was lifted, the need for cable TV would cease to exist. Skeptics likened the infant industry to the Pony Express—of momentary usefulness in the ocean of time.

"I can remember when our system had a three-channel capacity," says Sandford Randolph, who began his cable career in the early 1950s with the Clarksburg, West Virginia, system. "The afternoon telecasts consisted of a disc jockey who played records. He would introduce a song and say a little about the artist. Then the camera would focus in on an aquarium. So we watched the fish swim around while a song played."

"It was an interesting group—the cable pioneers," comments

Jack Crosby, who launched his cable career in 1954, when he built a cable system serving the Rio Grande community of Del Rio, Texas. "It was a cross-section of engineers, lawyers, goat ranchers, TV set dealers, Italian aviators, country squires, naval officers, Golden Gloves boxers, promoters, and others. . . . Believe it or not, very few in the group were clairvoyant."

In its early days, the community antenna television, or CATV as it became known, was indeed populated by a mix of visionaries, entrepreneurs, dealmakers, and others not so easily categorized. For many of them, just like the ones on the outside looking in, cable television was no more than a mechanical means of improving reception for TV owners tucked away behind interfering mountains or in rural areas too distant to pick up the weak signals transmitted from large cities. For those selling television sets, of course, the thinking was sound: by 1950, cable systems were operating in 70 communities and serving approximately 14,000 subscribers.

PROFILE

Bill Daniels

One man who believed in cable's potential beyond clearer reception and fish-bowl programming was Bill Daniels, who remains one of the industry's most oft-quoted protagonists—and one of the prime shapers of an industry that has grown. For thirty years he has been cable television's dealmaker extraordinaire, evangelist, apostle to the regulatory agencies, lead bushwhacker to the financial community, and one of the industry's founding fathers. Nobody has worked as hard for cable TV as the chairman of Daniels & Associates, the Denver-based concern that has served as the broker or arranged financing for more than $4 billion worth of cable TV property sales. In 1981, the firm did about 80 percent of the cable deals in the United States.

Simply put, there would be no cable television as we know it today without the street-smart, ex-Golden Gloves boxer named Bill Daniels. He is a former NCTA president and cable TV's champion promoter. Said one associate, "If Bill looked across the street and saw a glass of water, then, by God, that was the best glass of water in the world." It wasn't until 1952 that Bill Daniels saw television. Now it is said that Daniels did for cable what John D. Rockefeller did for oil.

I was living in Casper, Wyoming, and my mother lived in Hobbs, New Mexico, and we didn't have TV yet in either of those places. One day, on the way home from Hobbs, I stopped in Murphy's Restaurant on South Broadway in Denver. Actually it was a bar-restaurant and I've always been a bar-stopper. Well, they had the Wednesday Night Fights on in the bar, and I was completely enthralled! I didn't believe what I was seeing. I was thirty-two and it was unbelievable. The other day, I was in my office watching a fight from Atlantic City and I still didn't believe it. I guess if I was raised on TV cartoons, it wouldn't faze me at all. But it still does. I ended up going back to Murphy's every Wednesday and Friday night—for the fights on TV.

When was the first time you heard the words "cable television"?
It was also 1952. I talked to an engineer in Denver, trying to figure out if it made sense to build a TV station in Casper, Wyoming. He thought it was silly as hell. I figured there *had* to be a way to get signals into a small town. I mean, we were three hundred miles from the closest station. One day, on a trip back east, I heard about something called community antenna television, which I thought was a real hokey name. When I came back west, I wanted to make it happen in Casper. So I sold my insurance business to devote full time to it.

At the time, Denver had four television stations on the air, but it was so goddamned expensive to start a CATV system that we had just a one-channel capacity. I used a microwave, and that microwave pumped in the channel. We started out with five hundred subscribers in Casper, and I remember that we used to poll our customers to see which shows or which of the four channels they wanted to watch. We could only give them one, but they could choose out of the four. We'd mail ballots every week and I think "I Love Lucy" won a lot of the weeks. "I Love Lucy" was a big hit in Casper, thanks to cable TV.

As the Casper system flourished, you began buying other cable systems and putting together deals for other people. All of a sudden, people began paying attention to cable television?

All of a sudden, the *opposition* was beginning to build. Christ, I've been thrown out of more banks than any man in the country. They threw me out of banks in Casper, too. Oh, don't get me wrong: they'd lend you money if you personally endorsed the note, which was impossible. No one had that sort of collateral. So, for all intent and purpose, there was no financing available because the whole thing was brand new. Actually, they all thought I was nuts in the early days. Now I can look back and say that there hasn't been a bank that's lost a dime in a cable loan.

We had enemies from day one, that's for sure. Three of them were called NBC, CBS, and ABC. Particularly ABC. They used to call us thieves, yelling that we got our product for nothing. Now, of course, ABC is in the cable business. The local TV broadcasters were against us, too. Movie theaters were against us. So were movie producers, AT&T, 95 percent of Washington lawyers, a whole bunch of lobbyists, congressmen and senators, too. They thought we were crazy. Cities thought we were nuts, as well as the local, state and county branches of government. My attitude at the time was that none was more serious than the other, but it did tell me one thing: If our industry was not "onto something," then these people wouldn't have given a damn. All of them were trying to protect their own selfish interests.

What's your reaction when you see the commercial networks hooking up in cable TV ventures? Is there a sense of vindication?

Nobody likes to hear anyone else say, "I told you so," but I spotted cable before they did. I'm enjoying that they're in cable now. For years I beat my head against the wall, and they used to cut up our industry unmercifully. I guess they finally saw the light, you see. I guess the greatest obstacle to deals—and the greatest resistance to cable—came because of ego. I remember that CBS initially refused my suggestion to buy Ted Turner's network, and then CBS executives finally flew to Atlanta to negotiate, mainly because they didn't want Turner to be seen around CBS headquarters in New York. Isn't that stupid, when there are millions of dollars involved? But that was their attitude about cable TV; people couldn't be seen with them or there might be talk. Anyway, Turner and CBS negotiated for a couple of hours and didn't strike a deal, but CBS did swallow their egos.

Is there still room in cable for the young Ted Turners—or Bill Daniels?

I tell the kids—start as a janitor if you have to, just to get in the door. Cable television is growing, and there's a growing appetite for young talent. Personally, I never graduated from college and I didn't even know what an MBA was until a few years ago. What I know is how to make a deal. I have accountant types around me now, who give me the figures and I shove them in my briefcase and I fly off in my jet. I think the business is exciting as hell for anyone in it. It's exciting as hell for those who watch it. My advice to people who want to get in instead of just watch? Be optimistic, because you're dead without it. And observe the turtle: He progresses only when he has his neck out.

For the rest of the fifties, CATV companies were content to serve as alternatives to rooftop antennas, ensuring clear, ghost-free reception of local signals. Growth was steady, though not spectacular. By 1960, 750,000 Americans were fine-tuning with CATV, a figure that would double by 1965.

The relationship between CATV operators and broadcasters

soured as the 1960s dawned, and microwave transmission like the one Bill Daniels used in Casper, Wyoming, became more feasible from a technological and economic standpoint. Channel capacities on cable systems went from two and three, to five, to twelve, to twenty and more. High-powered, highly directional microwave signals meant distant TV stations could be pulled in with a minimum of effort—and operators began turning their attention away from "community antenna" toward the cable itself.

Operators of TV stations in the smaller markets feared that the importation of distant signals could siphon away their viewers and possibly put them out of business. CATV had begun attracting other Chicken Littles, too. Theater owners had become certifiable paranoids about cable TV. They thought cable TV would effectively shut down their business. Antenna manufacturers had something to say about cable, too. They formed a lobbying group called the Television Accessory Manufacturing Institute (TAME sounded better than TAMI), in an effort to halt the virus threatening to eliminate rooftop antennas.

Fueled by the support from the theater owners and antenna makers, the broadcasters took their case to the Federal Communications Commission. Throughout the 1950s the FCC had been reluctant to intrude on the CATV picture, reasoning that its domain over the broadcast media did not extend to CATV operators who merely passed signals over a wire, and not through the air. The FCC *did* have authority over microwave transmitting and receiving systems, however, and that technology was to serve as the basis of the first federal regulation of the cable industry.

In 1962, Carter Mountain Transmission Corp. of Wyoming asked the FCC for permission to import distant signals by microwave into Riverton, Wyoming—for the purpose of rebroadcast over a cable system. Riverton TV Station KWRB objected, claiming that the proposed system would effectively kill off its business. The FCC denied Carter Mountain's application because of the objections of the Riverton station, and Carter Mountain sued. The case finally reached the Supreme Court, which did not hear it but stated that the FCC clearly had the right to turn down an application if it would interfere with another service regulated by the FCC. The Commission defended its position against Carter Mountain—and cable TV—by determining that bringing additional television channels to the community and making them available via a cable system would result in the economic ruin of the local television stations.

In 1965, the FCC took its first step toward a comprehensive cable policy by issuing a set of relatively mild rules for cable systems that received signals by microwave. Less than one year later, however, the Commission lost all its shyness about cable regulation when it put into effect regulations that covered *all* cable systems, whether they used microwave or not. The most difficult to swallow, from the cable industry's point of view, was a requirement that cable systems in the one hundred largest television markets must apply to the FCC for permission to carry any distant signal. If any objections were raised by a local TV station to a cable system's plan, permission would be denied until hearings had been held.

Between 1966 and 1968, when the Supreme Court again upheld the FCC's authority over cable (in the *U.S.* vs. *Southwestern Cable* case, when the Court ruled that a San Diego TV station was harmed by the cable company's transmission of signals from Los Angeles), literally hundreds of applications for importing signals found their way into the FCC offices in Washington, D.C. The flood of applications was so great, in fact, that the FCC called a halt to the entire process. The cable industry was frozen in its tracks by the Commission's delay.

The FCC rules—and then its decision to stop and reexamine the situation—review the growth of cable particularly in the cities, where 87 percent of American viewers live. Part of the reason the cable industry slowed to a crawl was improvements in broadcast technology; in most places, cable was no longer the only way to receive clear TV signals. But the larger issue was variety, or more precisely, the lack of it—thanks to the FCC restrictions. Cable television simply didn't offer enough until it could offer program services that could not be had through broadcast stations in the same area.

Finally, after years of study, public hearings, and endless false starts, the FCC in 1972 issued its Cable Television Report and Order. In addition to mandating access channels (a rule that was abandoned in 1979 after the Supreme Court ruled, in *Midwest Video Cable Corp* vs. *FCC*, that the FCC was exceeding its authority), the Commission in that year thawed the cable freeze at least a few degrees by permitting the importation of three distant signals in large markets, two in middle-sized markets, and one in small markets. This was in accordance with the FCC's estimated ability of these markets to withstand additional distant signal competition.

At the same time, however, the FCC ruled that certain imported programs had to be "blacked out" on cable if they were also on local stations. The FCC set a requirement, too, that cable systems located within major television markets were to have an actual or potential capacity of twenty channels for broadcast and nonbroadcast service, in addition to the capability of two-way communications.

These last two rules were tossed out in 1979 when the Supreme Court ruled that the FCC was again overstepping its boundaries. Cable operators are now free, as well, to carry an unlimited number of distant signals and are no longer required to provide syndicated exclusivity protection. With the gradual elimination of the federal regulations governing cable systems, the doors were finally open for cable TV to mature past its role as a signal enhancer.

The first big step in an exciting new direction occurred on the evening of November 8, 1972, when some 365 homes in John Walson's Wilkes-Barre, Pennsylvania, cable system received the first "Green Channel" feed by terrestrial microwave transmission: an NHL hockey game followed by the movie *Sometimes a Great Notion*. A great notion, indeed. The Green Channel was to become Home Box Office.

At the time, the brand-new venture—subscription pay television—was nothing more than a pile of projections on the desk of Charles ("Chuck") Dolan, who in 1961 had founded New York City's first cable television service, which came to be called Sterling Cable. Dolan had the distinction of being the first person to build a cable system in a major American city—and he nearly went broke in the process. Delays, caused by bitter opposition from broadcasters and the telephone company, drove the construction costs beyond all projections. "The banks wouldn't lend us any money unless you gave them collateral," said Dolan, echoing the cable operator's plight in the growing phase. "They had no faith in cable TV because it had no great track record. The banks weren't alone. No one had any faith in cable TV."

Dolan had experienced rejection in the cable industry before. He originally had the idea of wiring New York City hotels in order for patrons to watch local conventions, but failed to inspire enough interest from an advertising community that had never heard of cable TV. "It was a great idea," said Dolan, "but a lousy business."

Once he had acquired the Manhattan-Cable franchise, which actually covered but a tip of the borough, Dolan attempted to put together a sports and movie package for his subscribers. This first move was to convince Madison Square Garden executives—and then his own investors—that putting sports on the cable was a logical start toward establishing local identity. He convinced everyone that it should be done, but the impact on subscribers was far less than what he had hoped for. "Somehow, though, it was still the right thing to do," he said. "Cable TV was growing up."

The next part of the package caused considerable problems for Dolan. At the time, he was prohibited by the franchise rules from showing motion pictures on his cable system—a stipulation that took up all of one line in the original charter. Dolan asked the city to amend the franchise so he could include films in his package to subscribers. Dolan recalls:

"It led to a full-scale debate with all the interests that opposed it. The theater owners opposed it—they lobbied heavily against it. Back then, you couldn't go into a theater in New York without being handed a petition to sign that said, 'Save Free TV!' I had this one little cable company in one small part of the city, and they mounted a full-scale, widespread effort to stop me from showing movies on my system. When it finally came down to the city's Board of Estimate—and the day they were to make their decision —all the theaters closed down and there were hundreds of theater employees picketing at City Hall. Thankfully, I wasn't recognized as I walked into the building.

"The city ignored all the pickets and voted unanimously to let us show movies on the cable system," he continued. "I think the first movie we showed on Sterling Cable was one part of a film trilogy from Satyajit Ray of India. The *Apu Trilogy* it was called, I think. Soon, we were showing films like *King Kong, The Naked and the Dead, 400 Blows*—films with a real sense of quality. A lot of distributors wouldn't make films available to us, but the ones we had looked so good, we decided to syndicate it. Hence the Sterling Movie Channel—or the Green Channel. "Still," he adds with a grin, "if we had paid attention to the surveys when we started it, we probably wouldn't have started it. The surveys indicated there was some interest in pay-TV, but not a whole lot. The problem was also the cable operators; we originally planned to have a Green Channel West, a Green Channel East, and a Green Channel Mid-west—with movies and regional sports delivered by microwave to those different areas. We didn't think of satellite at the time. Still, going back to the early seventies, many cable operators thought the next wave wasn't subscription TV at all. There was a vast indifference to the concept. They used to say—the cable business is okay; we improve the reception of broadcast TV. Why do something different? Why take a risk?"

Due to the astronomical costs of building his New York City cable franchise, and the costs of putting together his movie and sports channel, Dolan was forced to raise capital by going public; before long he had a new corporate partner, Time Inc. With each new public offering, Time worked to amass the controlling interest in Sterling.

By 1971, Time executives dominated the Sterling board of directors, and Dolan began to find himself on the losing side of nearly every vote. With some reluctance, he agreed to turn over operations to Time-appointed managers while he continued to work full time on his idea. He soon hired Gerald Levin, a young Wall Street lawyer, as the first program director. "I was the first," said Levin later, "because no one else would take the job."

By the end of the first year, 1,395 customers in Wilkes-Barre, Pennsylvania, were subscribing to a pay-television service retitled HBO. During 1973, HBO actually started to lose subscribers, before other systems were brought into the microwave fold. The Time Inc.–controlled Manhattan Cable system introduced HBO in 1974 (it took two years before Manhattan was capable of carrying the optional channel), and that year closed with 57,715 pay subscribers on 42 systems in 4 states.

The next seminal development was the decision by Time Inc.— at the suggestion of Gerald Levin—to gamble $7.5 million on a five-year contract to put HBO on RCA satellite SATCOM I, even before the "bird" was launched into geosynchronous orbit 22,000 miles above the equator. After several demonstrations of the feasibility of satellite delivery, HBO commenced regular transmission of its programs via satellite on September 30, 1975.

The FCC, of course, had something to say. Concerned that subscription programming on cable systems—"pay-cable"— would result in a loss of conventional television service to the public, the FCC adopted rules in 1975 restricting cable carriage of feature films, series, and sports events. These restrictions were challenged, and in 1977, the U.S. Supreme Court in *Home Box Office* vs. *FCC* affirmed a lower court decision that there was no demonstrated need for the regulations. The decision, coupled with the FCC's December 1976 approval of relatively inexpensive earth station antennas (which the cable operators could utilize to receive HBO programming off the satellite) led to the rapid growth of HBO affiliates and subscribers.

By 1977, HBO went into the black. Today, it numbers over 13 million subscribers. If Wall Street analysts are correct, HBO will have earned over $100 million in 1982, a profit from the networking of video entertainment that is exceeded only by ABC and CBS. On a dollar basis, HBO now has to be counted as one of the three big networks; it's doing better for parent Time Inc. than NBC is for parent RCA.

Chuck Dolan, meanwhile, sold his few remaining shares in Sterling Manhattan Cable to Time Inc. for $600,000 and made an offer to buy the Long Island franchise that Time Inc. had built. Today, Chuck Dolan's Cablevision Systems is the largest independent cable concern in the nation, operating franchises in Long Island, Connecticut, Boston, and Chicago.

"There were a few times, in the early days, that I thought I might quit trying," says Dolan. "But I thought that if I went back home to Ohio, I might not get a job." He pauses.

"For the past twenty years, it always seemed that cable was an inevitability, and my belief has been strengthened. Maybe I was too optimistic too soon. But I believe the wire entering into the home is going to provide the universe. I'm proud to be one of those who caught the future."

Home Box Office has changed the way we watch television. And Chuck Dolan, after nearly twenty years of flirting with bankruptcy, can finally afford to have fun in cable TV. By some estimates, he will be worth $400 million at the end of the decade.

The advent of the satellite age brought another phenemenon to life: the so-called superstation, which garners a national audience thanks to satellite delivery.

Before SATCOM, the broadcast range of Ted Turner's Channel 17 in Atlanta, for example, had been forty-five miles on a good day. On December 17, 1976, a little more than a year after HBO, Turner initiated the original "superstation"—he began to beam his station's signal to the satellite full time. The satellite instantly increased the channel's coverage to well over 10 million square miles, and today WTBS out of Atlanta is touted by its videosyncratic boss as the "Fourth Network"; it reaches more than 20 million households throughout the fifty states. Thus, the curious phenomenon of Atlanta Braves fans in Alaska developed.

As Turner tells it, the history of Turner Broadcasting, of which he himself owns 87 percent, is a testimony to the great man's tenacity and foresight. "If Christopher Columbus had a Southern accent," he declares, "then I'd be the man." Besides Columbus, he compares himself to Galileo, Robin Hood, Jiminy Cricket, and William S. Paley, founder of CBS-TV. "This Paley guy sounds kind of interesting," says Turner. "Maybe we ought to have lunch sometime. But it can't be right away, because I'm busy as hell."

The "Fourth Network"—or "Great American alternative," as Turner sometimes calls it—includes such television chestnuts as "Leave It to Beaver," "The Munsters," and "Gomer Pyle" as a hefty portion of its daily programming fare.

"Hell, there's nothing wrong with 'Gomer Pyle'!" says Turner, sometimes called "The Mouth of the South." " 'Gomer Pyle' is pro-social! The typical network mentality is to be number one in the ratings regardless of what you have to do, and that's why so much sex, violence, antisocial behavior, and stupidity has taken over the networks. The networks should put a disclaimer on their shows, saying, 'Watching this is dangerous to your mental health.' They look at the viewer the same way a slaughterhouse looks at its pigs and cattle. They sell them by the pound to the advertiser —the same way they sell ham hocks and spare ribs. They're worse than the Mafia!"

But is the great alternative called cable television really Gutenberg or just more glut? Fortunately, reruns of Herman Munster don't comprise the vast majority of the programming efforts of an industry still feeling its oats. The "made-for-cable" networks, made possible by the growing number of satellites hovering above, represent a broad mix of interests. For culture lovers, there's the ARTS Channel and Chuck Dolan's Bravo; ESPN (Entertainment & Sports Programming) offers the sports addict a twenty-four-hour-a-day video fix; Spanish-speaking movie buffs can tune in on Galavision; the Weather Channel means never having to say you're wet. Then there's the Playboy Channel for those who need a change of pace from Bravo; Ted Turner's Cable News Network for those obsessed with keeping up to date; the Satellite News Channel for information buffs who don't like Turner's; the Nashville Network for those who like their cable programming homespun; the Cable Health Network, which should advise you to stay twelve feet from your screen; a glorious array of religious networks; and others such as Showtime and the Movie Channel and USA and Daytime. There's even a network called C-Span, which is overwrought with congressmen, and Nickelodeon, which caters to kids who couldn't care less about Tip O'Neill. Add to that a network called Eros, which shows you things you didn't learn in school, and the Learning Channel, which shows you things you did. The Disney Channel is up, along with a skyful of others. There are plans for a few more cable networks, too. One of them, called the Legal News Network, could do for lawyers what MTV has done for dopers.

The cable explosion that had been in the promising stage for so long has begun. In 1979, more than 15 million homes were wired for cable, a figure that jumped to 33 million by the end of 1983. Some experts, schooled in cable optimism, predict that the industry will reach 80 million homes by the end of the eighties.

And the feast has not been prepared for viewers alone. Today's employment picture is beginning to reflect the rapid growth of the cable business. In 1975, fewer than 13,000 people were working in cable TV on the system level; today, that number has nearly quintupled. It will continue to increase as more cities are franchised and wired for cable, and new and better mousetraps meet the satellite age.

"Back when I started," says John Walson of Mahanoy City, "cable was just an idea. It was just a project. No one could have envisioned what it has become."

It took a few years, but the lesson is clear—almost as clear as the reception was on Walson's storefront TVs. Never underestimate the power of an idea whose time has come.

THREE

How You Can Become a Cable TV Superstar: Local Access

Try this simple quiz:

Local-access cable TV is:

(a) a sinkhole of moral depravity and a breeding ground of decadence, pornography, and people who wear leather just for the fun of it.

(b) an exciting arena for concerned citizens and nonprofit community groups to discuss issues of local concern at minimal cost.

(c) a learning environment for would-be television producers and stars, and an excellent starting point for a career in cable TV.

(d) all of the above.

Baffled? So are the people who produce, perform, pontificate about, and promote local-access programming on cable systems around the wired land. On the one hand, the media has done its best to publicize the seamier side of cable access, with frequent stories about so-called talent like "Ugly George" Urban, whose Manhattan-based cable program—"The Ugly George Hour of Truth, Sex and Violence"—did for cable television what a recent botulism case did for vichyssoise soup. On the other hand, there are literally thousands of unsung access-users producing alternatives to traditional TV fare, ranging from the deadly somber to the occasionally scatological to the charmingly off-beat. The fact is that local access has restored to the television screen some qualities that have been nearly refined out of it: spontaneity, originality, controversy, and realism.

In Madison, Wisconsin, access-watchers have learned from the "Oriental Art of Bonsai" show. Subscribers in Arlington, Virginia—just across the Potomac from Washington, D.C.—can see "Congressional Wrestling," produced by a young lawyer. In Los Angeles, a mother and daughter team dress up as rock stars, play games in their living room, and primp for the camera in a local-access hit. A local-access *series* in South Dakota shows the finer points of polishing wood. And on Manhattan Cable's Channel D, "The Waste Meat News" features the Leather Weatherman and a girl in a bathing suit who is tied up laterally to represent the map of the USA. The weatherman dramatizes snow and sleet by spraying her with whipped cream, dousing her with ice and water; he rounds off the weather report with gusts of wind from a blow dryer.

Then, of course, there's the access gem out of Georgia starring a woman who insists that Jesus is alive and well and working a night shift in downtown Atlanta.

The local-access movement is anarchic and often amateurish, struggling along in relative obscurity until one of its more inspired members drops his drawers on camera or threatens to blow up the White House. Yet the concept of providing local-access cable TV to community members is a rarity in a world where, as A.J. Liebling said, freedom of the press belongs only to those who own one. Access TV allows—or should allow—unfettered mass control of the media, instead of the other way around. In local-access cable TV, the mass media can become everyman's media.

Local access is basically free-form television in which people are welcome to do their own thing, even if their thing puts *you* to sleep. This independence suggests a peculiarly pure notion of free speech —the only glimmer of such a notion, in fact, in the Electronic Age. Peeved about the high cost of Mallomars? Interested in talking about your new religious cult? Anxious to tell the world—or at least your slice of it—about your latest idea called beef soda? There are currently close to six thousand cable systems across the country and nearly 50 percent of them offer at least one "access" channel. In some of the suburbs and cities where cable is new, many of the local cable companies offer *several* access channels, millions of dollars worth of equipment, full-time staffers to help you use it, and even mobile vans— all for community use.

The reason for this windfall of goodwill is simple: offering local access is a sure way for a cable company to snuggle up on the good side of the local governments, which make decisions as to which of the bidding cable companies gets the franchise in each area. Another obvious benefit to the cable company is that local-access programming helps differentiate the cable service from the competition, such as an STV system in town or the local independent stations. A recent study conducted by a national research firm revealed that potential cable subscribers list news, movies, sports, *and* local-access programming as among the major lures of cable TV.

A less obvious benefit to the cable company offering local access is that it provides a long-term talent pool from which the company can draw to meet the ever-increasing demand for people experienced in cable TV. Employees, interns, volunteers, and local-access producers and performers are gaining valuable hands-on experience through the opportunities of access TV.

Tom Cantrell, now a producer for Warner Amex's Qube system in Houston, used to work in local access in Austin, Texas. "The cable industry is growing at an incredible rate," he says, "and if the cable operator wants to grow with it, he'll need a ready pool of top-notch talent to meet his personnel needs. Local access is an exceptional training ground for developing cable-conscious future employees. You can find former Austin access-users working in cable from Boston to California. And they assume roles from production assistants to vice presidents. The fact is that people who work in access have to be an eager, hardworking, and resilient group. All in all, just the kind of people major cable companies want working on their team!"

"Access TV is a resource that's hardly been tapped," adds another voice. "I think it provides a great opportunity for the cable operator, the industry at large, and the people who want to get in. It's the one thing that separates cable TV from all the others."

The voice belongs to Fred Silverman, the former network honcho now tuned in to cable television. Think about it, Fred. When it comes to local access, you can't be canceled!

There are three similar-sounding but different types of access availability that fall under the general heading of local access.

1. *Leased access.* Many cable systems offer the opportunity for independent producers to lease portions of channels—or an entire channel—for the purpose of airing programs with a commercial theme or programs with a sponsor. The programmer—the buyer of the time slot—then takes responsibility for content, advertising, and cost. Leased channels are usually used by corporations that have an extended message to get across. Probably the most visible newcomer to channel leasing is the newspaper; approximately thirty-five daily papers have begun to program both character-generated and videotaped material onto the neighborhood cable system.

Leased channels don't come cheaply: on Los Angeles Group W Cable system, for instance, one hour of leased-access time costs $400. For most would-be cable stars and producers, there has to be a better way.

2. *Local origination (LO).* The cable operator owns and retains editorial control over the specially designated LO channels. According to the National Cable Television Association's latest survey, more than 70 percent of the systems responding reported that they offer their subscribers LO programming. More and more cable operators are realizing that they can earn additional revenue —and goodwill—by programming local-interest shows and selling the allotted ad time. LO is an opportunity for would-be cable

producers and stars to turn out—in association with the cable system—community-oriented programming. The more LO programming a cable system does, the more job opportunities, too.

3. *Public Access.* These channels, when available, are different because they belong to the citizens of the community according to the terms stipulated in the cable company franchise. Public access belongs to the amateur videophile in a Detroit suburb who produces a dog obedience school-of-the-air, and to artists, senior citizens, religious groups, librarians, cops, rock'n'rollers, exhibitionists, political ax-grinders, Indian chiefs, heterosexuals, MBAs, belly dancers, and anybody who has anything to say. It also belongs to those who have nothing to say except to their cat. It's television for the people, by the people.

The first governmental unit to establish access as a right was the New York City Board of Estimate. In 1970, the board awarded franchises to two cable companies, and numerous educational, artistic, and community groups developed very strong ideas about what "public access" should be. These groups banded together to help draft the city's franchise charter into the most liberal and clearly defined in the nation.

In 1972, the FCC launched public access by requiring cable systems in the top one hundred markets to earmark channels for government, educational groups, and the public. The FCC revised and softened its access requirements in 1976, and then, in 1979, the U.S. Supreme Court struck down all FCC rules requiring access by a vote of 6 to 3. The Court's decision was the culmination of a suit brought by the Midwest Video Cable Corporation against the FCC. Midwest Video, a cable operator, objected vigorously to the FCC mandate concerning access channels. It argued that the Commission had no right to impose access requirements. Midwest won; cable systems are not required by federal law to provide access channels. Today, individual states and localities decide how much, if any, access the cable company must offer. Many cable operators, especially those with limited twelve-channel systems, have long viewed public access as a public abscess. The reasoning was simple and based partly on the bottom line: Why give away a channel for free when you can fill up the space with programming that you can sell?

Over the last couple of years, quite a few cable operators—prodded by the community members—have agreed, ever so grudgingly, that access programming is generating positive results. "If you're *on* cable TV," says Sue Miller Buske, head of the National Federation of Local Programmers, "you will be more apt to buy it."

And if you're lucky enough to live in or near one of the towns where public access is offered, then you might find free studio time, state-of-the-art equipment, and an audience waiting for your show. In many towns across America, this grabbag of opportunity is all yours for the asking.

THE LOCAL-ACCESS JOB FILE

Access Coordinator   This person is hired by the cable system and is responsible for the operation of channels available for use by the public. The access coordinator, a position that is unique to cable TV, must involve the community in local access, and often conducts education campaigns to acquaint the public with the availability and operation of the service. If studio facilities are provided by the cable system, the access coordinator will supervise, assist, and monitor the programs turned out by the public. Thus, he or she must be a good *facilitator,* and provide the necessary training and tools so that people can develop their own creations.

Scheduling of access programs also falls under the access coordinator's jurisdiction. This requires making certain that there are no scheduling conflicts and that shows are aired on the access channels at the appropriate time. In order to become a full-fledged, full-salaried access coordinator, a minimum of two years' experience in all aspects of TV production is required, in addition to good writing and verbal skills, strong organization skills, the willingness to work flexible hours, a good sense of humor (you'll need it), and the patience of a saint. Remember: the access coordinator works with the public, most of whom have never met a TV camera face to face. The effective access coordinator will encourage the cable rookie, as opposed to breaking the camera over the good citizen's skull. Annual salary estimate: $12,000–25,000. For more information, see the profile of Ross Rowe (p. 40), plus Chapter 7 on An Education in Cable TV.

Assistant Director   It is the job of the assistant director to prepare the studio for production. This entails placement of equipment and assuring that everything is operational. The assistant director also coordinates timing with the camera people (e.g., cueing in commercial breaks), relays the director's call during production, and works at the video control board. Annual salary estimate: $10,000–20,000.

Audio Technician   This person is responsible for quality sound transmission in a production, which includes timing of musical cues and the monitoring of all sound effects for pitch and volume. The audio technician also synchronizes the microphones used during a program production. Salary estimate: $4–7 per hour.

Director of Local Origination   This cable system employee determines the needs of the community either through studies conducted by the marketing staff or through self-generated studies. After determining the community needs, the director supervises all phases of the system's LO programming, from the inception of the idea and writing of script through the technical production. This is a management position that requires supervision of all programming staff, hiring for the department, and overall budgeting for programming needs. Annual salary estimate: $15,000–30,000.

Editor   After the completion of the filming or taping, the editor, under the supervision of the producer, electronically splices the tape and produces the program in its final form. Editing may involve working with the artist/actor to ensure that the desired interpretation is presented. Salary estimate: $4–8 per hour.

Floor Manager During a production, the floor manager is in contact with the control room. He/She aids in camera cues and provides silent signals to the actors. The floor manager may also be called upon to keep a running time on the show and to make sure that it is sticking to schedule. Salary estimate: $4–8 per hour.

Intern Many local-access departments, at the system or community access center, are frightfully short on manpower, but rarely lacking in esprit de corps. Interns are chosen from local colleges, high schools, and street corners, and often end up operating cameras, directing access programs, and delving into virtually all phases of production. Although the internships are non-paying, an access intern is in an enviable spot. Interns are given a broad range of responsibility and they work on shows which are seen by the public—a marvelous incentive in lieu of pay. By nature of his/her position on the floor, the intern will also be privy to hearing of any job openings within the cable system. Said Barbara Shaw of Grass Roots Community TV in Aspen, Colorado: "I had absolutely no backgrond in TV when I started interning in 1979. But once I got involved, I was hooked! Now, there are former Grass Roots cable interns working full-time in television in Indianapolis, New York —all over the country. They all got their training in access TV." Barbara Shaw, the former intern, is now the one and only full-time salaried employee at Grass Roots Community TV.

Lighting Technician Shadows, highlights, and other lighting techniques provide overall impressions in a production, and can play a crucial role in establishing a scene. The lighting technician is responsible for reinforcing the mood and atmosphere of a production by creating special effects through lighting. Salary estimate: $4–7 per hour.

Producer   When a cable system produces its own programming, a staff must be provided for its operation. The producer of LO programming selects the cast and schedules the day-to-day rehearsals and taping sessions. The producer may also write the script or provide creative input into the script, as well as provide the visual interpretation for use by the camera and video staff. In smaller operations, the producer may also work the video control board. In some small systems, producers are contracted for a single production; in larger systems, they may be part of the permanent staff. When it comes to the public-access channel that a system might offer, the producer is not only the creator of the show idea but often the writer and star as well.

Unlike their counterparts in title who produce for the major cable networks, would-be access producers do not have to worry about creating shows with commercial appeal. Commercial appeal and local access seem a contradiction in terms. However, the access producer must think up projects that have a neighborhood flavor and feel, especially if the producer wants funding from the local system. If funding is needed—and the show could make the cable system look good in the eyes of the community (i.e., potential subscribers and city council)—then it is possible that the director of local origination might see fit to hand over the system equipment and maybe a few dollars in cash to help the show get made. If you have an idea that will only interest your gerbil, then the odds are you'll have to pay for the space yourself. That's the real difference between local origination and public access. In LO, the cable system is the producer's partner. In public access, the cable system will probably charge a small fee for the space.

No matter how you plan to utilize local access, the range of available program topics is virtually limitless. Producers in local access can create shows on:

1. *Politics and local issues.* Interviews with candidates for municipal and county offices are popular offerings at the local-access level. So are interviews with the local mayor, city council people,

and other political animals who live down the street. The first local-access show ever was an interview with the mayor of Pottsville, Pennsylvania. One of the early believers in cable TV, Martin F. Malarkey, did the interview for his own cable system in 1951.

Documentaries about minorities, the aged, and abused children often are seen on the local-access channels, especially the ones that feature LO programming. The show must capture the street beat of the town; Viacom's cable system in San Francisco offers "Lovestyles" on its Channel 6. It is a weekly half-hour show dealing with the concerns of the gay community, and is co-produced by the cable system and a San Francisco resident.

2. *Public service.* Even the most reluctant of cable operators have a place in their pocket for community members who want to do the community good. If you come up with a public-service idea that would bring tears to Mother Theresa, you might shake some coin from the local cable Scrooge. One of the recent local-access award winners came from Greendale, Wisconsin, and was called "Diaper Dippers." The program was taped at the local YMCA, and pinpointed a popular infant swimming program in the community. The producer of that show was a genius: absolutely no one wants a baby to drown. Other award-winning access programs have offered tips on health care, fire and crime prevention, and helping the handicapped cope.

3. *Oral history.* Often produced in conjunction with universities and historical societies, this genre documents a community's past. Cable systems across the country are encouraging a sense of community heritage. A producer in California's Marin County came up with a documentary on the last manned lighthouse in the region, at Point Bonita. Interviews with retired lighthouse operators and their families brought to life this soon-to-be-extinct occupation. The dedication and isolation of these people were captured in the faces and words of people who seemed to have jumped out of a Grant Wood painting. Another show, produced for the Sioux Falls cable system in South Dakota, captured the reminiscences of a one-time homesteader, an immigrant who had survived winter

storms and deadly disease to preserve his land. And a third, "Irving Weber's Iowa City," features the local historian and his tales about the city's landmarks and historical figures. The show is seen on Hawkeye Cablevision in Iowa City. Irving Weber is eighty-two.

4. *Sports.* The networks and local indie stations have most of the sports scene sewn up, but local cable channels can and do provide coverage of high school games, college events, and special community sports outings. Cox Cable in Oklahoma City cablecasts fifteen Oklahoma City Stars hockey games. "We were really excited about being able to keep up with the puck," says the director of LO, Joan Young. "We've gotten as good a reaction to hockey as we did from the rodeo." Producers in Oklahoma City help her put the sportscasts on the cable. Another cable system tapes the annual "Bedpost Races," a local event during which citizens push beds down the street; on the beds are other citizens who do their best not to fall asleep.

5. *Entertainment.* Durham Cablevision in North Carolina puts on a local producer's show called "The Channel 86 News Fest," which is a merciless send-up of the inanity of local news shows, complete with fumbling sportscaster, prop-happy weatherman (no leather here), and a recent emergency report of an impending "meltdown" in town. It turned out to be a story about a Good Humor truck with a malfunctioning freezer. Santa Barbara Cable TV worked in a show that parodied the community's Fiesta Week with a happily malicious account of the *real* story of the town's traditions and real estate development. In almost every town that offers local access, there is a show with a predictably amateurish Johnny Carson, a wonderfully bad would-be crooner, and a wooden dancer who clippety-clops all over the channel, trying to woo and win the big-time Joe who might be in West Virginia from them thar mountains of Beverly Hills. There's a producer behind each and every one of those shows, either hiding or trying to promote it. Annual salary estimate for a producer employed by the cable system: $10,000–25,000. For more insight, see the profile of Penny Nathanson (p. 44).

Production Assistant   The PA assists whoever he/she is assigned to during a production. A production assistant may take notes during a runthrough of the show, take and transcribe notes for the assistant director, Xerox production material, run errands, help move props during a rehearsal, answer the phone in the studio. The job can be a nonsalaried intern position or a salaried position on an hourly basis. Salary estimate: $3.50–5.00 per hour.

Studio Manager   He/She is responsible for the overall technical management of the studio. This manager—or technician, if you will—must repair broken equipment, diagnose problems, purchase equipment, and ensure that the facility is operational. The studio manager will also pitch in and help with the scheduling and traffic in the studio, and may be called upon to be a camera person at times, too. Annual salary estimate: $15,000–25,000.

Star   There is no other way, for some, to feel truly self-fulfilled unless he/she can get on television. Don't worry that William Paley doesn't return your calls. Forget the fact that you have no discernible talent. Local-access cable TV means that anybody, or almost anybody, can have a television show to call their very own. It's like those old Mickey Rooney movies: "Hey, kids, let's put on a show!"

Not everyone wants to end up on a shelf at the Museum of Broadcasting, of course. Not everyone is obsessed with making it to "the big time," a place, we suspect, that is much smaller once you get there. For many, the grass-roots movement of the people's TV—local-access cable—succeeds just because it's used; as a hobby, a caprice, or a way of reaching out. There is something to be said—hell, there's something to be cheered—for just enjoying yourself. With all due respect to Ugly George, thousands of people across the cablescape think local access is the most fun you can have without taking off your clothes.

And if you do want to become Johnny Carson, local access is the best place to launch your face. Since no one really expects cable-access programming to be slick and professional, you have a golden chance to test your wares. In the wire mire of local access, if you can smile without drooling down your chin, you've scored a few points already. What really matters is the quality of your invention and imagination. You will have a better chance of making it if you let yourself be you, instead of a bad imitation of your favorite star. If it's inspiration you're lacking, remember that Robin Williams got his start on a local-access channel in star-struck L.A. It was on access channel 3 that the pre-Mork Williams made his television debut in 1975.

For more information on the local-access stars of today, see the profile of Mark Gauger (p. 54) and the next section on How to Be a Cable Superstar.

PROFILE

Ross Rowe
Access Coordinator

Local access has been a fact of life in East Lansing, Michigan, since 1973, when volunteers from Michigan State University recorded school meetings and events with a black and white "portapac" and the system ran the tapes on the "weatherscan" channel, replacing the weather with the videotapes from the school. Local access grew through tales of begging and borrowing equipment in order to tape wherever the original access-users had an interest. By 1974, East Lansing's local access was operating out of the basement of the University Inn (a local hotel next door to the cable company's future office building site), with one coordinator checking out por-tapacs and arranging editing time on the prehistoric equipment.

Now, with a budget of $35,000 for capital expenses and $80,000 for operations, the access channel has a name, WELM-TV, and two studios (1,500 by 850 sq. ft.), a mobile van, editing facilities for Betamax and ¾-inch cassette, four color portable cameras, and five color cameras for in-studio use.

WELM—cable-access channel 11—currently airs thirty-five to forty hours of programming per week, and it has been estimated that eight hundred people use the facilities monthly. Program diversity is apparent; "Uncle Ernie" narrating his vintage home movies, Sloucho Barx hosting "Tee Vee Trivia," and Michigan State ice hockey are among the access hits in East Lansing. Ross Rowe, the access coordinator, is responsible for tying it all together. He runs the workshops that are offered to community members, covering basic television production functions. During the seven two-hour sessions, Rowe allows the participants to get valuable experience handling the cameras, the lights, and the buttons in the control room. Rowe is also responsible for stimulating interest in the access program, and making sure that the access-user is well prepared when he or she does a show. At twenty-seven years of age, he is one of cable television's new breed of workers; he came upon cable television as a career quite by accident:

I was going to school at Michigan State; I had transferred there from Wayne State in Detroit. I was taking a few video productions in school, learning the ropes a bit, and I knew nil about cable TV. I didn't know that local access existed. This was about four years ago; and I had a job working part time at Sears. I wasn't too excited about the possibility of working full time at Sears, either. About this time, I moved in with a new roommate who had cable TV, and I was amazed. What a wonder! I can see M*A*S*H five times in two hours!

How did watching cable for the first time lead to your job in it?

One night, I was switching around and I zipped right past an ad on a cable channel. I zipped right back to it and the ad said: "Do you want to be a TV producer? Come down to WELM and get involved." Well, I knew that whenever they use the words "get involved," it's gonna be volunteer work. "Get involved" and "work for free" go together. But I thought I'd keep my hand in TV production while I was still in school, so I went down, took one of their workshops, and volunteered.

My first assignment was lead cameraman for a women's basketball game. The hard part there was keeping up with the ball. Then I was exposed to all different types of programs. And I got involved in all aspects of production. Everywhere else but in cable TV, you're told to do one thing in production—and one thing only. Don't touch anything else! In local access, I could do—and I was asked to do—just about everything. So, little by little, I got into the access bit; but my heart was still set on big-time broadcasting.

What was your next step?

I got a part-time job working with the East Lansing school system: we put on programs at night that aired on the cable system's school channel. The educational channel. Then, one day, I saw this little announcement in the local paper: "Needed: local-access coordinator at United Cable." People used to call it tinker-toy television. They said the equipment wasn't up to snuff. But I decided it would be a real challenge. I also needed the money. And they hired me, mainly because they knew me from all the volunteer work I had done. They remembered me as the guy who could keep track of the ball. I'm not making the big bucks that some people make in broadcasting, but I'm enjoying myself. I still get to do everything!

It isn't a nine to five job, that's for sure. If you're looking for a nine to five job in cable TV, go into insurance. In fact, if you want a regular life at all, don't get into this business. I leave my house when it's dark in the morning and I come home when it's dark at night. I haven't seen my house in daylight since I started here. The studio is open from 10:00 A.M. to 11:00 P.M. every day, and we usually never go out for lunch. And there's always paperwork to do before the studio opens and after it's closed. In fact, I'm not so sure I have a house, anyway.

If I wanted to do my own access show in East Lansing, I would have to see Ross Rowe first. What would you tell me about the requirements?

The only real requirement is that you're an East Lansing or Meridian Town resident. There's no age requirement—if you're under eighteen, just have your dad take the equipment out. We offer workshops, but I think you lose some of the spontaneity of access when you *require* the workshops. So here I might say, "Okay, here's how you use the equipment, let's run through it once or twice." Then I'll let you take it out. *Go for it, Ace!* That's our motto here.

Of course, if you break something, you have to take financial responsibility. If you drop the portapac out of an airplane just to see what it tapes on its way down, you're buying us a new portapac. That's in the agreement you sign. The only other cost involved is if you want to make more than one copy of your tape. It costs ten bucks per dub [copy].

Are you glad that you're here instead of working at Sears—or possibly CBS?

I never liked the feeling of "Okay, here's your eight bucks an hour, thanks for coming." Here, I get to work on the whole she-bang. And I think the whole shebang is getting bigger and more important. Cable TV is where it's at.

From what I've heard, there are a lot of jobs being made available in local access, too. In the Chicago area, someone told me that the Warner Amex franchise hired a total of fifty people to do local production. Welcome to the club: fifty more people who will never go out to lunch!

PROFILE

Penny Nathanson
Access Producer

Vision Cable in New Jersey's Bergen County is working with a local producer on cable television's first full-length, made-for-access *motion picture*. The 25,000-subscriber system and the producer Penny Nathanson are taping local talent and volunteers for the project, which will offer a realistic look at the day-to-day problems and lifestyles in the New Jersey area. No one ever promised that it would play like *Gandhi*. Twenty-seven-year-old Penny Nathanson does insist, however, that the local-origination movie will use "all-Jersey actors, all-Jersey crews, and all-Jersey jokes."

It's gonna be a series of vignettes. The script, which I wrote, calls for sixteen leading roles and about fifty or sixty extras. I want people to say, when they see it, "Hey, that's my mother over there!" I want them to see houses they recognize. I want them to say, "That's my life." Personally, I don't relate to Jane Fonda at all. I'm not sure that there's such a thing as a Jane Fonda, anyway. So we took the Hollywood out of it. Even the musical score was composed by a localite, a seventeen-year-old cable subscriber. And they say New Jersey has no talent outside of Springsteen!

Who says that?
I don't know. Springsteen's mother, maybe?

How much is the made-for-access movie going to cost?
It's very very low budget. It's a beg, borrow, steal, volunteer-type movie. The locations that we're using *are* people's homes. Sometimes, we tape in someone's home for fifteen hours and they ask us not to come back. That's a problem, when you have to find

another house. A few times, the microphones blew, too. We have had all sorts of technical problems. But I can't estimate how little it's costing all told. I would venture to say that it's the lowest-budget movie in the history of the earth. The cable company is picking up the tab, of course. And I get a weekly paycheck while we're doing, but it's really low, too. Once it's over, I have to pound the pavements again looking for a job. At least for a couple of months, I'm getting paid and gaining experience. For a couple of months, I'm in the movie biz!

How did you get the idea to produce a made-for-access movie? Have you been speaking with Dino de Laurentiis?

Dino eats breakfast with the budget of this movie. But I always wanted to do a movie ever since I was a kid. The project actually started when someone at the Vision Cable staff thought of producing a talent show to showcase local entertainers. I went to them and said, "You're right. Cable should be community-oriented. Now let's do something different!" I told them I'd write the movie and produce it, and I guess they had no one better so they went with me. I was elected to do everything. I think what really sold me was my line: "Hey, think of the PR!"

I had worked for Vision Cable for about four years, then I wanted to try freelancing, so I left. But, as I found out, freelancing is tough, even with four years behind me at the cable company and a degree in psychology. When the movie idea came up, and it was really a fantasy of mine, a lot of my friends were still at Vision Cable. So I had an edge into the company. I'm not sure I could have convinced them to do it without the contacts that I still had. I knew the right people in the right places. Even if I didn't, though, I would've given it a try. I think I have guts—in a stupid kind of way.

One of the dumb things that I did was major in business and psychology. I mean, I wanted to be a newscaster and producer. But one of the bright things that I did was volunteer for Vision Cable at the start; I worked for the first three or four months without pay.

After volunteering, I got a regular job with the company.

One of the worst things to have in this business is a lack of direction. Vision Cable has intern programs and part-timers and a lot of them come in and don't know what they want to do. And some of them even have communications degrees! I think it's important to find out what you want to do before they have a chance to let you go. Narrow in on your talents. That's what I did. I asked myself: "What am I good at? Can I BS a little? Can I work a video machine?" Do a little soul-searching and find out who you are. Then sell yourself. I may be out of a job in a couple of months, but I know one thing: I'm a producer. I just finished a movie! You can bet that fact will be underlined when I send out my next résumé.

HOW TO BE A CABLE SUPERSTAR

If local access is available in and about your hometown, it is vitally important to know how to go about getting your fair share. In other words, there are rules of the game, and if you don't know them, you might as well try for a career in robotics.

Case in point: Donald Swann of Kankakee, Illinois. A couple of summers ago, Donald Swann wanted to be a television producer and star, and he intended to start on a local cable system's access channel. Donald Swann gathered up his savings, bought a second-hand video camera and recorder, and began taping his program in and around Kankakee.

He didn't want to produce the run-of-the mill access show, such as an interview with the mayor's wife. He wanted to make a documentary special all his own. So he began spending a lot of time in and around other people's bathrooms. He had conceived of a show called "The John in Society," and felt that the time was right for a visual inquisition. As the host of the special, Swann believed it would garner him much attention in Illinois and possibly land him a spot on Merv. He was only half right.

Donald Swann zeroed in with his camera, and occasionally

caught a surprised subject with his or her pants down. No Ugly Duckling this Donald Swann, he would excuse himself, back out of the bathroom with his camera, wait for the person to finish the business at hand, and then proceed, at the wash basin, to ask appropriate cinéma vérité–style questions! "Do you read in the john? Do you have to run the water in order to go? Do you always leave the bathroom door unlocked? Have you ever squeezed the Charmin?"

Swann put it all together in a half-hour show, did a taped intro-duction which he managed by setting the camera on a tripod, and contacted his local cable system. Since Swann had never contacted the cable company about his plans, they knew nothing about a show featuring area bathrooms and the social significance of porce-lain seats. Because the mere mention of the word "bathroom" offended "community standards", Donald Swann was told to look elsewhere for a place on cable TV.

He finally gained access to a local-access channel some miles away. The cable system scheduled his program—without looking at it first—and Swann made plans to return to the town when it aired the following Friday.

That Friday, he found a bar in town that had cable and con-vinced the bartender to switch from ESPN, the all-sports channel, to his show. Immediately, Swann knew something was wrong. His opening shot of a ladies restroom had been cut. So had the inter-view with his neighbor, who had reminisced about missing the bowl after a big night of downing Bud. The quick shot of the gal toilet-training her baby son was also nowhere to be found. He watched his half-hour show end after sixteen minutes, and then Donald Swann walked the twelve blocks to the cable company office. By the time he got there, he was steaming mad, and asked, in a tone that was hardly conversational, to see the director of local programs. He also asked for his copy of the tape.

When both arrived at the reception desk, Swann fired off a few choice invectives, yelled something about freedom of speech, grabbed the tape, reached for his blue Bic lighter, and tried to set the tape aflame. A few sparks nicked some papers on the desk and they started to go up in smoke, too. By the time the director of local programs had sprayed the area with the cable company's extinguisher, it seemed half the desk was burning and the fire was threatening to spread to a cushioned chair nearby. A security guard had arrived on the scene and Swann was hustled away. The burned papers and the smoking tape were kept as evidence.

Last time anyone checked, Donald Swann had soured on cable TV, and cable TV had soured on him. There *is* a moral to this story, of course. First, always make a copy of your tape. Secondly, know the ins and outs of the local-access business. Since the rules and regulations can vary from system to system, it's important to know what you're getting into.

And finally, never ask a bartender to switch from the all-sports channel.

If there is an open channel or two set aside for access on your local cable system, and you can overcome any lingering technophobia, the next step is to know how to start your career as a star in cable TV.

Pop a dime in the phone and call the cable company (it's listed under Television in your Yellow Pages). Ask for the director of local origination, the access coordinator, the studio manager, or anyone involved with local access. You should know that many cable companies are notoriously slow in picking up the phone. Hang in for ten or thirty rings. Take a deep breath, but don't hold it if they put you on hold. Finally, you'll get a chance to speak to the person in charge.

If you're interested in doing a show, which is presumably why you called in the first place, you should know that there's a chance that no access is offered. This is particularly true of many of the older cable systems, which have a limited twelve-channel capacity. However, a 1983 NFLCP survey identified over 1,300 systems nationwide which offer access in some shape or form, and over 800 access-management centers operated by community groups. Virtually all of the new-builds in Dallas, Boston, Cincinnati, Chicago and elsewhere are offering various kinds of local access like they really mean it. Cablevision of Boston, for example, is offering five public-access channels, and the Boston Community Access Programming Foundation will be operational in early 1984, with four studios, training programs, and personnel looking to involve the public in access. A spokesman said that close to twenty people will be hired to run the access center, and there will be an increasing need for interns and volunteers.

In Dallas, the Warner Amex franchise will offer sixteen access channels, which is about as serious a commitment as a company can make.

Since there are no federal regulations governing access programming, there are no set and steadfast rules in terms of cost. In Tidewater, Virginia, for example, the system offers five minutes free of charge, but if you want an hour, it will cost up to $600. In Baton Rouge, Louisiana, you'll have to dig fairly deep into your own pocket, too. The cable system there charges $216 for a one-time slot on the so-called access channel 8. If you have a weekly series in mind in Baton Rouge, it will cost $140 per half-hour slot. There is no studio or equipment available through the cable system, either. "And we don't run anything controversial," says Bo Berenger, a system official. "We're in the Baptist Belt here and if we ran something even mildly controversial, we'd be nailed by the local DA. He watches our shows real closely—and he'd come looking for us."

In San Antonio, Texas, there is no cost at all for filling time on Channel 21, but there's a long list of people trying to get on, so you have to wait your turn. "As long as the program contains no profane language, it's all right here," says Sharon Mooney, the access coordinator. "It can't have any nudity, either, of course. And," she adds, "if it's really controversial, I don't know what we'd do with it. We had a person come in not too long ago with a tape about Communists. It dealt with something about Communists being behind attacks in the area. He had no facts to back it

up so we didn't air it." Dimension Cable (Times Mirror) in Louisville, Kentucky, offers two hours of free studio time for each half-
hour show, and charges no admission to the access channel. The
system will even bicycle your tape to the Storer Cable system,
which serves a different chunk of town. The one catch is that the
system will provide a director, but no camera operators. "You
have to bring your own," said Sue Ann Lyons, the traffic manager.
"But if you bring a camera person who has never touched a camera, we'll provide a free training session. Within an hour, he'll be
zooming, panning out, dollying, and doing what a camera person
does."

In Bloomington, Indiana, Don R. Smith runs a community-
access channel, which is leased to him by the cable company for
$1 a year. He programs seven days a week on the cable's Channel
3, and turns out five to ten new access programs per week. All of
it—from the channel space to the workshops to the videotapes—
is offered free of charge.

Access channel 12 on the Group W Cable system in Los Angeles
is on the air every night from seven to midnight. There is no charge
for the use of the channel, but fees are applied for the production
or prescreening of access shows. If a show is not produced at
Group W's studio, there is a fee of $15 per half hour. This is known
as the "handling fee." For a show produced at the cable company's
studio, the cost is $35 per half hour. This covers what is known as
"the setting up of the chairs"—and the actual production of the
program. Access channel 12 on the Canyon Cable system in Aspen,
Colorado, is run by the Grass Roots community group, which
trains cable ingenues and provides equipment free of charge. The
non-profit organization has been in operation since the fall of 1971,
when it began production of public access programming with a
wide shot of Ajax Mountain. The shot of the mountain comprised
much of the Channel 12 fare for a couple of years. Today, the Grass
Roots movement, funded through the community and an annual
telethon, produces original access programming ten hours a day,
seven days a week.

If local access is offered in your hometown, the next step should be the workshop. Chances are that either the cable company or the local community-access group runs the workshop, and it will serve you well to spend a couple of hours learning what a TV studio looks like and how a TV camera zeroes in on your face. These workshops, like the one offered in Louisville, are designed for those with little or no practical experience in TV production.

Most of the workshops are offered free of charge to interested community members. In Reading, Pennsylvania, the Berks Community TV Group, which produces fifteen hours of live access programming *daily,* runs training workshops and also has volunteer technicians available to help community members who want a bit more expertise. In San Francisco, Viacom Cable offers free workshops three to four times a week, although the system may begin charging a nominal fee to cover equipment costs. In Seattle, Washington, Group W runs a workshop designed for potential users of access channel 18. They charge a $10 material fee for the workshop. San Diego's Cox Cable system, the largest in the country in number of subscribers, offers several different video workshops at $40 each.

If your local cable contact has an access channel and a workshop to boot, the next step for the soon-to-be cable star is to fill out the application. Before you get to use an access channel, it is likely that the cable system will send you an agreement that defines in basic terms the rights and responsibilities of the applicant and the cable company itself. At the United Cable system in East Lansing, for instance, the access-user must sign an agreement which states that the user assumes all liability for program content and agrees to indemnify the cable company, the city of East Lansing, the Charter Township of Meridian, and its representatives for all liability or other injury due to program content. That's rule 3.2 under the user-responsibility clause of the agreement. Nearly all the applications from around cable country have a similar rule. It means that if you call your neighbor a "local putz" on the air, and he decides to sue, you are the party responsible. The cable company wants no part of your problem, and it's only fair on the systems where the cable operator assumes no editorial control.

Most applications also clearly state that the access-user—and cable star—can't cablecast any advertising or lottery information, nor any indecent or obscene material. Most cable systems have contracts with the individual cities and towns that refer to "acceptable community standards." Remember Donald Swann? Some systems also add language in their applications/agreements that might not be seen in other parts of the country. In the applications of a number of cable companies in the Southwest, "anything that calls into question the divinity of Jesus Christ" is strictly prohibited. And in East Lansing, there's a clause that prohibits an access user from drinking soda pop in the control room.

After the initial call, the workshop, and the inking of the application, you will be ready to enter the world of the "people's TV." It has been estimated that no more than 6 percent of community members participate if a local-access channel is available. Thus, there is bound to be a considerable amount of time open to you.

The next question: What kind of show do you want to do?

"Access channels on cable systems can be the twentieth-century equivalent of open space and parks," says Ellen Stern Harris, vice president of the Public Access Producers Association in Los Angeles (PAPA). "Access channels, as well as parks, provide the spaces people need: to explore, to exchange ideas with people they've never met before, to share their experiences and talents—to let one another know what's happening in the neighborhood and the region. People have got to know that they have an 'in' to community television. It can work for anybody who wants to give it a try." Gene Bell, who runs a small grocery store in Muscatine, Iowa, hosts a public-access show called "Local Color." I think it's a good idea to give it the old trial and error," he said. "I started about fifteen months ago, and I wanted to interview people who do outrageous things. I talked to a guy who kept snakes in his wall, for instance. I wear a cowboy hat so I can put people at ease. Hell, they say, if he's wearing a cowboy hat, I guess it's O.K. to talk to him!" Bell's show is cablecast four times a week—free of charge—and he has become a local hero in Muscatine. "Nowadays, I get recognized on the street! Hey, people say, I see you on cable TV! I know you! The other day two girls asked for my autograph." Concludes Bell: "The key thing is to do a show that lets you be as good as you can be. If you can play the guitar, do an instructional show. If you like to speak to people with snakes, do a talk show. Access to us is the place to be! You can be anything you want to be!"

Plugging into the access channel with your very own show, as Gene Bell did, is easier than you might think. A recent seminar, moderated by Ellen Stern Harris, included tips on how to put together your first cable program. The underlining theme was that you didn't have to look like Tom Selleck in order for cable access to work:

- Outline the show's concept and theme—what you want to say, how and to whom you want to say it—on index cards, and then stick to your video game plan.

- Select a format. Will a talk show, comedy, lecture, how-to demonstration, musical or dramatic sketch be the best way to present your ideas? You might want to ask yourself: How does my show differ from the programs already on the access channel? Also, will you tape in studio or go out on location?

- Write a basic script. It should include the story, characters, and guidelines for set design. Listing items from backdrops and music is advisable. If you plan to do a talk show, figure out who you'd like to have as a guest. Then write out a dozen questions which might bring out the personality in that guest.

- Outline post-production needs, from editing tape to taping music. Checking copyright rules for borrowed music or illustrations could save legal hassles later.

- Budgeting is essential. As you know, some cable systems charge a nominal fee for tape stock, studio rental, and/or channel space. Others offer the services and space for free.

- You might also want to give some thought to promoting your show. Even though it's on the access channel, that doesn't mean that people can't be convinced to watch. In Iowa City, for instance, a group of performers formed Iowa City Underground Television and produce a show called "Space Heaters," which airs on Hawkeye Cablevision's Channel 26. The group prints up fliers about the show and distributes them throughout town. And every time the show is cablecast, they invite people to the Hilltop Tap Bar to watch it on a big screen. They usually draw between eighty to one hundred people per showing.

You should consider sending announcements to special-interest groups in your area that might be in sync with your show. Some cable systems have a channel reserved for public-service announcements and will run your ad for free. The chain call is also a possibility; call several friends and ask them to call a bunch of people announcing the program's airdate. You might also, if you're especially proud of your idea or the show after it's finished, send some information to the critic at the local newspaper. That's what the "Space Heaters" people did, and they got a glowing review in the Iowa City paper.

PROFILE

Mark Gauger
Iowa City Cable Star

Iowa City Underground Television is based in Incognito, Iowa, and is comprised of several young men who produce and star in a local-access gem called "Space Heaters"—a series that chronicles the adventures of intergalactic traveling salesmen. Given its financial limitations (funds for the production of the show depend on nickel can returns and garage sales), "Space Heaters" is really no worse than what passes for comedy on network TV.

More important than the product is the spirit behind it. The fact that "Space Heaters" can be done—that six people in Iowa City can have fun with television—is one of the great liberating blessings of local-access cable TV. Mark Gauger describes how it came about:

A bunch of us started out running an underground free-form FM radio station when we were in college. That's how we all got together. We used to do special features on the radio like "The John Denver Death Hoax," "Benny Benzadrine for Campus Drugs," and other socially enlightened stories. We were a few college students having a good time. When it became apparent a few years later that access to TV equipment was there—and time was available to us on the local cable—we just about pissed our collective pants!

When we got together to do the first "Space Heaters" show on access, the first episode was written by committee, which is like playing soccer with ideas. On the other hand, directing by committee is like getting a pregnant hippo out of a mud wallow—and we've done that, too. Critically viewing the first few episodes, I'd say they're crude and lack certain commercial standards. Some of it looks like it was taped on Alienvision. Our spaceship, for example, is called *EnRoute,* and it's made out of a bedpan. One of the guys, D. J. Beard, paid for the bedpan. The picture of Leonard Nimoy cost a nickel, which I got from returning a bottle. All the costumes we use are ex-Halloween costumes, which our cast made on their own. The equipment is the cable company's. If you exclude the beer we drink, it doesn't cost much at all.

What is the premise of "Space Heaters"?

It's basically this: a group of intergalactic traveling salesmen or salesbeings are going somewhere on this semi-spaceworthy vessel *EnRoute.* In the first episode, which was repeated on the cable a couple of times, a defective Part E burned out the delicate Blornge Sensing Unit. Blornge rhymes with orange, by the way. In search of a replacement Part E, the crew decides to beam or ream down to a planet, where they encounter people who keep telling them,

"Great costumes!" You see, they landed on earth and on Hallow-een. So the aliens reply, "Great costumes!" just to be friendly.

Where am I? Oh yeah. The aliens ask for Part E, and one of the earthlings says, "Heck, I just came from a helluva party!" So they go to the party, and for having the best costume, they are given Part E. In Episode 2, their ship passed through a zone of HiPi interference and most of them were zapped into Space Lawyers. Future episodes include: high times at the intergalactic truck stop and a show about Kephlan, the ship's navigator, who goes off to college and flunks astronomy.

Why are you doing a local-access show? What do you want to accomplish?

We're trying to brighten the vast wasteland of television and, by doing so, maybe help redefine what TV is—or could be. Actually, we don't plan to take the show anywhere. It's just enough that it's being done. In Dave Eifler's words (he's one of the group), when it stops being fun, we'll stop doing it. Otherwise, expect to find us down at the Hilltop Tap when we're sixty-four, trying to knock together Episode 608.

The fun that we're having is enough. We experiment and tinker, all in color on our hometown TV. I didn't have a TV for three years. When we started doing "Space Heaters," I finally broke down. D. J. Beard brought my dead Zenith back to life. And from doing the show, I find myself *watching* TV, looking at camera angles rather than gazing glazed at the tube. I feel more involved. I feel like I'm part of the process, instead of a passerby.

IS THERE LIFE AFTER ACCESS?

Can local-access channels serve as launching pads for independent producers and stars who want to wear Gucci shoes? Is there a way to make the cavernous jump from cable access to big-time TV? The answer is yes, but not a simple yes.

The odds are that if you want to produce another show like Ugly George's, you won't find a cable system in sight willing to air it—and certainly none around willing to buy it. Chances are, your mother will disown you as well.

It's no secret that the cable industry is rather sensitive about its access programming, and that many cable operators have tried to discourage their access-users from promoting shows, let alone syndicating them elsewhere. Undaunted, several have begun to achieve recognition beyond cable access and even gained a few dollars in the process.

David Jove, for instance, did an access show in Los Angeles called "New Wave Theater," which premiered on the cable channel 3 in early 1980. Currently, the show airs nationally on the USA cable network on Friday and Saturday nights; it can now be seen in over 15 million homes.

"My original intention was to get a show on the air, any way I could," says Jove. "Anything that is seen has a future. Exposure is the name of the game. Access TV, like new-wave music, is a real equality of experience. It's a great opportunity to be noticed." Jove ended up at a cable convention in 1981, where he showed some examples of his show. Then a company called ATI, which produces a four-hour block of programming entitled "Nightflight" for USA, called and made a deal for his show. It has been on national cable for close to two years.

Says Jove, "My original budget in access was about $1,000 a show, and now it's up to about $4,000 a show. I'm not making much money at all. But I'm hanging in, because this show has gotten me off in the right direction. And if you're not losing money in cable, you're winning." Jove continues to produce access shows, by the way. His latest, "Brown Box Theater," is aired every week in the same time slot, and on the same channel, that he started with three years ago.

And then there's Paul Ryan, who began with a local-access talk show in Los Angeles in June of 1977. Since then, he's hosted an NBC-TV show called "Singles Magazine," done various guest spots on network TV, and currently cablecasts his interview series over the national Satellite Program Network. "Cable access created an opportunity for me that didn't exist before," he said. . . . "I called up the system and said that I wanted to host a talk show, and they asked me, 'When do you want to start?' It was there for the asking and I seized it. And a lot of people started to watch me. Stars like Dudley Moore, Michael Caine, and Sophia Loren began to come on as guests."

And what did it take for Paul Ryan to make the leap from local access to a national audience?

"A lot of tenacity," he says. "You have to love what you're doing and you have to create something different. Local access is like a fantasy for most people. Hey, I can get on TV! In order to get bigger, though, you have to accept getting bigger a little bit at a time. And you have to get other people enthused with your own belief!"

Another access to affluence story is told by Arnie Rosenthal, who began his cable adventure on Manhattan's Channel J. "I was making documentary films in Soho for a place called Global TV," he recalls, "when a spokesman from Manhattan Cable came down and told us about plans for a thing called J, on which you could buy time and sell advertising for your show. I was never a great one for filling out job applications at ABC or going on interviews at CBS. I had $700 in the bank and thought I could produce a lot of shows for that amount. It started on March 2, 1976, and I was on with two shows—a full-time job. I moved back to my parent's house, and I lived off them. Occasionally, I'd get a short gig as a musician for a few bucks.

"One of the shows I did was called 'Jazz from Boomers,' which is now a defunct jazz club formerly of Manhattan. I did it in black and white, and on half-inch tape. The other show got me some notice; it was called 'The Big Giveaway,' and it was really the first two-way game show. People would call in on the phone, and it was really involving. And it got the ball rolling for me. I went door-to-door selling sponsors on the show, and I tried all sorts of tie-ins and gimmicks. I sold the show to local pizza parlors. I sold it to bars. I sold it to a barbershop, which gave us a free toupée to give away on the air. And I was getting sponsors that couldn't advertise on regular TV: ez wider rolling papers company was a sponsor on my show. By about 1977, I was doing 'Big Giveaway' shows on other cable systems in the New York area. I went from system to system. At this time, there was no ESPN, no MTV, no USA. The only cable network around was HBO. So there was a lot of channel space.

"About this time, one of the people who had helped me with 'The Big Giveaway' started getting more interested in cable and had the rights to a show called 'Pro Celebrity Golf,' hosted by Bing Crosby. I think it was the last show Crosby did. It had an American golf pro and an American celebrity versus a British golfer and British celebrity. Well, I thought the show could work in cable syndication, and I had a minor track record with 'The Big Giveaway,' so we biked it to one hundred systems, one by one. And since the golf shows were being produced every year, we thought we could use it as a kickoff to a package of foreign-produced shows. So we approached Granada, which is a European television company, and told them, 'You've got the product; we think we can sell it on cable TV.' I mean, they had twenty years of shows just sitting on a shelf. So we made a deal for their English programming, named our package 'The English Channel,' and picked sixty-five of the largest systems in the country for our syndication. Then, when we managed to get Volkswagen to buy four minutes of advertising time every night, it put us over the edge. All of a sudden, we had a cable network."

Nine months into "The English Channel," Rosenberg was contacted by Jean-Claude Baker, the producer of a public-access show in Manhattan called "TeleFrance." Baker wanted Rosenberg to syndicate "TeleFrance" much like he had the English shows. "I didn't know if it would play in Peoria, but I gave it a shot," says Rosenberg. "And it worked."

A European film company named Gaumont entered the picture

and supplied "TeleFrance" with enough money, and French-language programming, to go on the satellite. Rosenberg was asked to stay on and run the new TeleFrance network, which he did for three years. Now, he's the vice president of marketing and affiliates for Financial News Network and is a well-paid consultant in the cable industry. "Local access is the vaudeville of TV," he says. "It's a chance to break in your act. And if it works, you might have the chance to take it on the road. Where else but in cable TV could I have gone in with $700 and ended up with my own network?"

Still not convinced? Still worried that no one will watch you? Karen Salkin has been hosting "Karen's Restaurant Revue" on access channel 12 on the Group W cable system in Los Angeles since September 17, 1982. Karen has been called a restaurant critic, but it is very doubtful that she would ever risk eating for fear that it would slow down her speech. Salkin would much rather talk about her trip to the gynecologist and Disneyland, or her ear problems, or her dating tips for committed singles, or anything else she can convolute. Karen Salkin is the Princess of Piffle. She gabbles at such a rate that only dolphins can decipher.

Karen Salkin is a true blue access eccentric. "My boyfriend Ray watches TV incessantly. I mean, he has two TV sets on top of each other, so he doesn't miss anything. He said that I should do an access show. I thought they were all idiots, but I'm also an actress and I had an agent at the time who was a waitress at Kanter's. I mean, my agent couldn't afford batteries for her phone machine. So what I did for a living was throw dinner parties. Except I didn't cook, because my mother didn't cook, so I wasn't doing much at all."

Karen Salkin began appearing on public access "just to prove to my boyfriend," she said, "that nothing would happen." Something happened. "I was on my way to work out at Nautilus, and I stopped by my service to pick up my mail and phone messages. I read a message and it said 'Tonight Show called for a possible guest appearance.' Hah! I thought it was a joke. Everything's a joke to me. But when I called the number, it really was NBC, and I was told that the producer of 'The Tonight Show' watched my show and then I sent them a tape and I called back and they told me that Mr. Carson had the tape. A couple of days later, they called back to book me on the show. I almost died.

On July 22, 1983, Karen Salkin made her first appearance in the hot seat next to Carson's late-night throne. "I was a wreck," she recalled. "I saw him backstage and I want aaaaahhhh! I screamed! I went up to him and said 'I can't believe it's you!' At about the same time, I noticed my name on the dressing room door and it was spelled 'Darren Salkin!' "

During the show, Carson mentioned that he had seen Karen on access channel 12. Chevy Chase, a guest on the same program, mentioned during a commercial break that he watched her, too. For Karen Salkin, it was a night of dizzying highs. Her mouth went national.

And a couple of weeks later, Johnny Carson invited Karen back for a second shot on "The Tonight Show." "I still pay my thirty-five bucks to do my public-access show," she added. "I mean, I can't believe how much I've gained from doing it! First of all, I wanted to do the Johnny Carson show my whole life. And I've learned a lot, too. If I ever get a sitcom, I'll know all the stupid stuff, like what the red light on the camera means. When I first started my access show, I thought the red light was on to match the color of my dress!"

There are several more conventional ways for an inventive local-access show to achieve national exposure—and possibly a touch of stardom for the host and producer.

The Satellite Program Network (SPN)   Based in Tulsa, Oklahoma, SPN is a satellite-delivered service that reaches over 5 million cable homes across America. It calls itself a "lifestyle" network, which means it has a melange of programs ranging from the dreadfully bad to the decidedly mundane.

Satellite Program Network, in need of programming and sponsors to sustain its lifestyle, has aired several local-access winners on its service. Connie Martinson does a book review program on L.A.'s cable channel 12, which is also seen on SPN. The "Career Woman" series started out in local access in Memphis, Tennessee. If your show is approved for airing on SPN, you will still have to pay a fee to place it on the schedule. Currently, this fee ranges from $500 to $800 per half hour, and you have to be willing to commit for a minimum of thirteen weekly shows, preferably twenty-six. If you have produced a one-hour special, it will cost you $4,000 to cablecast it on the network.

Such fees are probably too steep for most access users, even those with the largest of dreams. But if you have a dream *and* a few potential sponsors who might cover your costs, the Satellite Program Network might make a bit of sense.

Marilyn Perry, for example, buys time on SPN to distribute her access show nationally, and she has an unusual way of financing it. "International Byline" features a different member nation of the United Nations each program, and the costs of production and syndication on SPN are picked up by the various nations. What this amounts to, of course, is an entire show dedicated to advertising the country that foots the bill. Ms. Perry might not be able to call herself objective, and she can't call herself rich, but she *can* say that her cable show is national. For some, that's worth the price of admission.

The process for getting on SPN is fairly simple: Call their Tulsa office and ask for Cherlyn Hampton, the programmer who got her job, incidentally, after reading a want ad from the network in a local paper. She will ask you to send a letter of introduction and a sample tape—or "pilot"—of your proposed series. "We have quite a few talking-head programs because they're the least expensive to do," she says. "So if you have another one, it's got to be different from the rest. Also, how-to's are doing very well for us. Our sewing show might not have the best production value, but it has a loyal audience."

In your letter of intro to SPN, detail whether you plan a weekly show or a daily one and underline the fact (or hopeful fact) that you have one or two potential sponsors ready to fuel your production. Mention, too, when the sponsors might be ready to come up with the cash in order for you to begin your national show.

"If we're interested in your program," says Ms. Hampton, "we'll check our schedule and see if we have a time spot you like. Then we'll phone you back and see if you're still interested in SPN. If we don't like your show," she adds, "we will still phone you back. And we'll return your tape, too."

The Cable Health Network   CHN works in quite a different way from SPN. Whereas SPN takes a rather scattershot approach in its programming, the Cable Health Network is a satellite service with a perspective and profile. It is a channel devoted twenty-four hours a day to helping people understand their bodies—strengths, weaknesses, joys, and mysteries.

CHN also differs from other cable networks in that imagination and ideas may be more important in getting started there than someone's level of experience. "I look for people with something diverse, something interesting, someone who uses his imagination," says Ron Ziskin, a West Coast executive with Cable Health. "I'd rather have ideas than a track record." Actually, they'd rather have a show in the can. CHN execs believe the opportunities are limitless for local producers, but stress that the technical standards must be up to par. Unlike SPN, you see, Cable Health Network *buys* programs from local producers. The money isn't much—it ranges from a couple of hundred bucks to a couple of thou—but to a producer with a show already done, the money is pure gravy. And the prestige of getting a program on CHN is virtually priceless to a young access producer.

Don't count your chickens before they cheep: Before submitting your program to the Cable Health Network, call its New York or Los Angeles office and ask for a package of information, which will contain a rundown of the programs currently on the network. These include shows on nutrition and diet, self-help and medical care, growing up and getting older. Use an ounce of common sense:

CHN is not interested in your home movies or your series about the history of lint.

Also pay particular notice to your production standards. "I've seen many access programs with excellent content," says Paul Coss, the vice president of acquisitions, "but I haven't been able to consider them because of the poor technical quality."

Coss offers some hints to the local-access producer on how to improve the quality of his shows. "Not enough attention is paid to the lighting of the programs," he says. "It often looks flat or there are too many shadows. A show has to be well lit to have that professional look. Sets also require more attention. Most often, they're unimaginative . . . a couple of chairs with a coffee table. Why not go out of the studio and use someone's home or office, for a more realistic atmosphere and more attractive surroundings? Or, if you must shoot in a local studio, consider working with a local furniture dealer who might be most willing to dress your set for an on-air credit.

"Sound quality is another area where improvements can be made," continues Coss. "As much attention should be paid to achieving top-quality sound as is given to the program's video look. As an example, I saw one excellent nutrition and exercise program from the Midwest where the audio was so poor that every time someone moved, the clunking of the microphone across their clothing could be heard. Or the performers would move out of audio range, making them inaudible.

"In considering product for acquisition," he concludes, "we must look at the total package—the excellence of the program content as well as high technical standards in both the audio and video portions of the show. Each aspect is important in terms of evaluating the material, and in determining its acceptability for use on Cable Health Network. You must realistically look at the need of the potential buyer."

If you have produced a local cable show that might fit the CHN profile—a show about breaking the smoking habit, for instance—you should follow the network's basic submission guidelines.

1. Before you submit a program, the network prefers that you send a letter describing it. This should include content description; when and where the program has been seen; and a bit of background about yourself.

2. Productions shot on half-inch tape are *not* acceptable. Half-inch tape is less expensive but fails to meet CHN's quality control standards. One inch is consistently dependable for high-quality recording, editing, and airing.

3. Cable Health Network prefers to screen ¾-inch cassettes of programs, although if necessary it will screen 16mm films.

Bicycling   Think of the success of Arnie Rosenthal. Think of the movie *Breaking Away*. That's the sort of upbeat yet underdog mentality it takes to make it happen through "bicycling" your show. Bicycling has nothing to do with Schwinn in this context, unless you want to cut down on your mailing costs.

"Bicycling" your show means distributing it by mail to cable systems outside your immediate area. In the long run, this mode of distribution is cheaper than securing time on SPN. In the short run, however, it will add up to a fairly hefty phone bill—and takes a good deal of hustling.

If you want to increase the coverage of your local cable show in the state where you live, the logical progression of events should happen like this: First, go to your local library and find out the names and numbers of the systems within state lines, or thereabout. Choose a few of those systems and call the local program managers. Most of them will be at least willing to listen to someone who might have an interesting piece of software. If your show is currently on a cable system in your hometown, throw in the names of a few people the program manager might recognize.

Unless you have something really marketable (such as an exclusive tape of Andy Kaufman wrestling Greta Garbo), the program manager won't pay cash for your show. However, if he has a leased-access channel or an LO availability, he might agree to cablecast it on a barter basis; he'll put it on for free if you allow him to sell one or two of the commercial minutes within the program. Don't haggle. The increased coverage will add to your own selling package. The key word is credibility. It takes more than luck to get a sponsor interested in cable TV.

If all goes well, the bicycle system should work as follows. Once you've gotten a yes from the first cable system on your route, a few more are apt to come along for the ride. This doesn't mean that you have to send different tapes to each of the systems along your "network." Bicycling only works if one cable system forwards it to the next, and the next to the next, and so on. The mailing costs may be billed back to you, but hopefully your advertising revenues will cover the administrative costs.

One local-access producer, Alan Orenstein, currently bicycles his show to fifteen different cable systems. He used to be on Satellite Program Network but dropped out when his sponsors became dubious of the network's real audience. "Now, with bicycling, I don't reach as many systems but I have more control," he says. "I even get to promote the show, since I'm within a car drive of most of the systems."

To a local sponsor, eight or ten cable systems concentrated in one basic area might be worth more than a few hundred systems scattered about the cablescape. And don't think that this concept has escaped the grasp of the cable operator, either. Several enterprising systems around the country are initiating "interconnect" plans, which entail the sharing of programming with other cable systems in their area. Advertising rates are then based on the aggregate number of subscribers, and the cable systems share the revenues. The thinking is sound: an advertiser would rather make one bigger buy than eighteen tiny ones. Plus, many of the regional interconnects feature local sports events, which advertisers seem to like.

And when it comes to local cable, advertisers usually don't like a lot.

GET THAT SPONSOR!

You've got a local-access series and, judging from the reaction in Podunk, you think it will be a national smash. Or your show "How to Play the French Horn" has knocked 'em dead in Franconia Notch.

You know enough to realize that most of the big advertising dollars in cable TV are being funneled into satellite-delivered networks like Getty Oil's ESPN—the twenty-four-hour-a-day sports channel. And you know that Cable News Network, the around-the-clock talking head propped up by newsmaker Turner, is attracting advertiser interest. So is his WTBS, the superstation; and the USA cable web; and other biggies—like Hearst-ABC's "Daytime" and the ARTS Channel. They're constantly on the hunt for Mad Ave's buck, and sometime they even get a few.

So where does that leave you with your half-baked cooking show and self-styled syndication?

Fighting the odds, that's where. On Satellite Program Network, for instance, you're allowed to sell five minutes' worth of commercial time every half hour. Pocket a few corporate backers, and it would seem easy as pie to arrive in the black. Remember: they charge you around $500 per half-hour slot. Peanuts to the Planters Company, right?

Right—and wrong. The large companies, through their large ad agencies, aren't buying big in cable TV. Especially on networks such as SPN. "I went to a big advertising agency in New York with a copy of the show I wanted to syndicate on their network," says one young public-access guru, "and they told me to bug off. They wouldn't even look at the show. Once they heard public access—and SPN—they all went to lunch or had their secretaries tell me they were in meetings. The place emptied out like I was a bad fart."

On the other hand, Nikki Haskell got a few local sponsors to back her show on Manhattan Cable—and then bought her way onto SPN. The increased coverage enabled her to raise her rates.

"I think my show projects smart, fun, bright things that are happening all over the world," she says. "It's really just a projection of my own life." Nikki, who calls herself the "cable baby of the eighties," had an "in" to the people with the dollars: her show seeks out the rich and powerful for interviews, usually in discos and other glittery and insoucient in spots.

Michael Irving also has had some success with sponsors. He produces a local show that is also seen on SPN, called "Traveler's World": "It's necessary to create a businesslike atmosphere," says Irving, "so that you can deal with your sponsors from a position of strength and confidence." Michael Irving owns a very businesslike English accent, and several very nice suits. That's important when it comes to selling yourself. If you don't have one, you should at least have the other.

The key strategy in snaring sponsors for your syndicated gem is selling them on the hopefully original qualities of your program. Special-interest programming—or "narrowcasting"—usually has a special-interest sponsor to match. It's like computer dating. One of cable's most successful programs was produced in Canada for American consumption. Called "Canada—News from Home," the series has played on SPN and speaks directly to Canadians living in the United States. Sponsors gearing their products for that very select Canadian audience, those who may have moved from Winnepeg to Wichita Falls, have been willing to pay as much as $2,000 for a single minute commercial. The series is operating in the black, which is more than you can say about Cable News Network.

If you're going the Madison Avenue route, it would make sense to evaluate your potential sponsors in relation to the program you're producing. Ralston-Purina probably won't jump at your idea for a series about new-wave music. Ralston-Purina might not be interested in your idea about pet care, either. When it comes to sponsors who are not gung-ho about cable from the start, a quote from Ms. Roseannadanna might be in order: "Oh well, it's *always* something."

In dealing with potential sponsors on any level—from Madison Avenue to the local sporting-goods shop at the corner—it's important to know what they're looking for. You can't point to ratings and shares, "the numbers," in cable television, at least not on SPN or the bicycle route. Godhead Nielsen has only done studies on such cable giants as Home Box Office and Ted Turner's networks. Without such vital info, the best you can do is point to the coverage maps of SPN, which are included in their package; or to the charts and graphs that you have drawn to indicate the coverage on your "network."

You can also try to sell the sponsor on the special qualities of the cable subscriber. Though this would get disagreement from some sources, the majority opinion among cable observers is that the demographics of basic cable homes are similar to those of noncable TV homes on average, *but* that those who pay extra to subscribe to pay services such as HBO and Showtime tend to be upscale in income and education and tend to be younger families with children.

There's one known factor that would get no argument: cable households spend more time watching television than noncable households.

If you're doing a series on the practical importance of teeth, you might convince a sponsor that the cable household earns more money, watches more TV, and flosses like mad.

ABC/Heart's "Daytime" network uses the pitch—derived from "a major study"—that the cable woman is better educated, shops a great deal, has an upwardly mobile career, reads a lot of magazines and books, and enjoys many leisure activities, and entertains more.

Doesn't sound like she has much time for "Daytime," does it?

Anyway, selling cable television to the advertiser large and small is a matter of enlightenment, wearing nice suits, lowering your expectations, and coming up with a show that is tailormade for the sponsor that you're pitching. It might also help to invest in an attaché case. Cramming your tact and figures into a Scott Baio lunchbox just doesn't cut it when you're on the wrong side of the desk.

NEW YORK, NEW YORK: A PROFILE IN ACCESS

Local access in New York City is the Jerry Brown of cable television. No one knows what to make of it. "They're all maniacs here," says Coca Crystal, star of local access's number-one anti-authoritarian show, "If I Can't Dance . . . You Can Keep Your Revolution." "Access in New York provides an opportunity for weird sick people to air their views."

Quixotic Coca says she doesn't quite remember her age. She *can* recall getting kidnapped in Bombay, taking acid in Frisco, being hijacked to Havana ("I asked them if I could stay"), and sitting on poll results she didn't agree with when she worked for Lou Harris.

Made on a shoestring budget, Coca Crystal's show is a five-year veteran of the local-access movement. In 1983 she was named one of cable television's top ten stars in a national publication. With tongue firmly in chic, Coca tosses her head back and says that she's writing a book called *Fame on a Budget.* A skimpy budget, indeed. She subsists partly on food stamps and pays for part of her studio time with credits earned as a volunteer. "To me, it's amazing that on the same box that you see Bob Hope, you can also see me," she says. "When you're on cable TV, you've got the power."

Coca produces her show at the ETC Workshop Studio on East 23rd Street, right next door to Manhattan Cable's headend. Back in 1974, when ex-ABC programming executive Jim Chladek wanted to do a live cable show in New York, he found that there was no place to originate his show. Undaunted, he rented the studio space, threw some wires out of the window, and ran them up to the cable company. Today, Jim Chladek runs two studios, and both are connected to the access channels available on Manhattan's two systems.

Some of the equipment at ETC might have been used by Ernie Kovacs, but since it can't be verified, the equipment is best described as old instead of historical. Once, the control room board caught on fire in the middle of Coca Crystal's show. The ventilation is provided by crew members who breathe heavily. Still, the low price (you can rent a studio for $20 an hour), the feeling of camaraderie and persecution, and its live capability attract many of New York's access stars to ETC.

Cranking out shows from the dusty studios is "The Crank Call Show," during which viewers are encouraged to call in and swear at the host. "It must be some kind of therapy," shrugged Chladek. Tex Fenster—star of "Tex Fenster, Superstar"—also makes it happen at ETC. Fenster is a middle-aged postal worker who lives with his mother in Queens and stars in a weekly program that, once in a while, finds him fondling magazine centerfolds on the air. "I'm a star," he says, "and I never even knew it." Morris Fonte the Telepsychic emanates from ETC. So does "The Spiritual Healing Show"—with Dimitros and Gregorian Dharna. They purport to heal people who touch the TV screen while they're on. Nicholas Yanni produces his authentic talk show from ETC. Chladek also hosts a show there; every Sunday night, he invites viewers to call in and talk about more serious subjects than Tex Fenster does.

In New York City—the number-one television market in the country—an access user can grab a chunk of free air time on Channels C and D. If a sponsor is ready to tag along, the producer/star can lease time on Channel J for about $150 per half hour.

"Cable-access people are way off the mainstream," said Jim Chladek. "They spin on their own radials. Some do it to make money and are failing now. Others do it to get laid, and they're succeeding. The fact is: they're doing it. All I know is that it's not the same yellow goo we see on commercial TV."

A. C. Nielsen would turn over in his plot. A series for test-tube adults? Necrophilia Tonite? The Waste Meat News? Who cares about the money, right? It's Vanity Video time.

CONTROVERSY IN CABLE ACCESS

Television kills.

A couple of Christmas nights ago, on Manhattan Cable Channel D, in living black and white, a small nondescript dog, tied to a stake, was being shot to death.

The thirty-second loop was repeated over and over, the same grisly scene: the dog, the stake, the gun, and the dog—crumpling to the ground.

Several weeks later, spokespeople from the cable company were squirming under TV lights they didn't own. It was all a big mistake, they said. Somehow, someway, the tape slipped through our hands. Actually, the cable people said, the system received only one call of complaint the night the show was seen. But the smell of death really hit the air when a local paper got wind and printed the DOG SHOT headline.

We love little doggies, the cable people insisted. We had no idea what was going on, they said. We're upset that it aired on Christmas night. In fact, one of the spokespeople added, we're even more upset that it was replayed a few nights later.

Ooops!

No one really planned to make it a mini-series. No one likes watching Snoopy look-a-likes get blown away, point blank. Especially around the holiday season.

The New York City cable franchise charter was drafted in 1970, and sought to protect that all-American institution "freedom of expression" by insisting that cable companies had to offer public-access channels and couldn't legally interfere with the programming. Today, cable companies are supposed to abide by "acceptable community standards"—which, in New York City's case, are as difficult to define as Andy Warhol's hair color. (Warhol and his hair can, of course, be seen, on public access in New York.)

Rule number 10 on Manhattan Cable's public-access handout sheet reads: "To assure maximum opportunity for freedom of expression by members of the public, programming on the public channels shall be free from any control by the Company as to program content. . . ."

The public-access channels, part of the basic cable service to all subscribers, are offered on a first-come, first-served basis to groups, nonprofit institutions, and whoever else feels the urge to communicate. Existing law forbids the New York cable operator to censor what is shown. Dismayed by the blatant sexuality on channels that were originally intended as outlets for free debate or showcases for fledgling performers, Assemblyman Melvin Zimmer of Syracuse pushed for a new law that would have authorized cable operators to screen out shows that contained "patently offensive violence, obscenity, indecency, or profanity."

The Zimmer bill amounted to overkill. Ugly George, who prowls the streets of New York and persuades women to strip in front of his camera, may be offensive to some viewers, but who can be trusted to define obscenity? Many people believe that the way women are portrayed on network TV is obscene. A fleeting reference to homosexuality or even the most decorous discussion of sex would inevitably offend someone.

The bill was defeated, but the question remains: Does a cable subscriber who wants protection against pornography in the home deserve to have it? Perhaps the appropriate response is to allow subscribers the power to screen out or limit access to offending programs. Instead of decreeing vague and dubious standards for censorship, perhaps cities should instruct cable companies to offer one access channel for clearly inoffensive programming and another channel where almost anything goes. A *New York Times* suggestion was to supply cable homes with devices that allowed the subscriber to lock out any channel, permanently or temporarily.

The practical and constitutional way to resolve the issue is to let subscribers become their own censors. Of course, that may make far too much sense for anybody ever to expect that it will happen.

Although programs such as Ugly George's, "Midnight Blue," and "Hot Legs" constitute but a small minority on local access, their disproportionate notoriety has made them rallying points for advocates of censorship—or cable company control.

Produced by former radio talk show host Alex Bennett and financed by *Screw* Magazine maven Al Goldstein, "Midnight Blue" has been on cable access since 1976, and, compared with many of the access shows around the country, looks as stylish as *Last Tango in Paris*.

Alex Bennett said he doesn't quite remember his age, but talked almost convincingly about his role as an access martyr. "I have a historical perspective," he said, between bites of a well-worn fingernail. "I want to be in the history books as the guy who opened the door for shows like 'Charlie's Angels.' "

Bennett likens his "Midnight Blue" to a sexual "60 Minutes." The program opens each week with a cowgirl riding a huge phallus over the Manhattan skyline. You immediately get the feeling that Morley Safer isn't in town. Right into your living room via Channel J comes bondage, boot-licking, and endless stories about local bordellos. Occasionally, Buck Henry shows up. Regular features include the sex survey (Is it bestiality if the goat loves you back?), sexual consumer guides, and video centerfolds. Once in a while, "Midnight Blue" covers record-breaking performances, although not of the variety that airs on ESPN. "We cover things as journalists," says Bennett. "We have dignity. Personally," he adds, "I don't think you can masturbate to 'Midnight Blue'."

For economic reasons, Bennett is now trying to promote the middle ground. Cable systems around the country are exhibiting an increasing appetite for R-rated programming, and are accepting such national, satellite services as the Playboy Channel and "Private Screenings." To keep pace, Bob Guccione of *Penthouse* has announced plans to launch a "Pet" Network.

But Bennett knows that if he is to have any success in syndicating his program around the country, he has to soften the blow of

"Blue's" reputation. "My show is sex without the headaches," he says. "It's not steamy, it's classy."

One of the first systems to carry Bennett's "Blue" outside of New York was the Perry Cable franchise in Stuart, Florida. Bennett had been denied time on the RCA satellite and had to take to peddling (or bicycling) his series of "Blue" programs system by system, state by state. He offered the programs at $250 each, with the understanding that the system leave in the commercials he had sold in New York.

From 1978 through 1980, the Stuart system had been offering "Midnight Blue," along with a variety of soft-core titles, to subscribers who had already bought Home Box Office. They offered it free of charge to the premium subscribers, in what might have been the best deal in cable history. Every night, after HBO went off the air, (this was before HBO went 24-hours-a day) the local cable operator in Stuart would slip on a sex feature. The response, needless to say, was overwhelming.

Still, all did not remain rosy pink in Stuart, Florida. Something went awry with the automation machine in the studio, and "Midnight Blue" popped on for five minutes at twelve o'clock. Noon. In stepped a vigilante group called Citizens Against Pornography to picket against the system. "It's a real shame," said then-program manager Bob Wilson, "but many cable operators are not in a position to stand up to these kind of groups."

A few weeks later, Perry Cable yanked "Midnight Blue" off the cable.

Said Alex Bennett, doing his best martyr: "I still think there's room in cable TV for 'Midnight Blue.' We might be targets, and we might be scapegoats, but this program will live on. Cable subscribers want sex!"

Compared with Ugly George, Alex Bennett has led a charmed life in cable TV. Not only has the irrepressible George Urban been denied time on cable systems around the country, he's actually been kicked off Manhattan Cable TV periodically over the past five years, for reasons too petty to describe here.

"The Ugly George Hour of Truth, Sex and Violence" has been called—and not without reason—the most pornographic program in television history. "He is sexist, manipulative, and antisexual. Worst of all, he's boring," says competitor Alex Bennett. "Truth, sex, and cellulite. Ugly George wallows in the swamp."

And he's a certifiable public-access legend. Believe it or not, his might be the best-known name in cable television outside of HBO and Ted Turner.

PROFILE

Ugly George

What George does exactly is the easiest and most difficult thing to describe. Hanging on a hook in the makeshift TV studio that is really the bedroom of his fourth-floor walk-up in Brooklyn (he moves just about every six months, and was last living in New Jersey) is a skimpy, tattered one-piece silver suit. Think of a skid-row spacesuit. When the weather is warm, George wears that suit on the streets of Manhattan. When the weather isn't, he wears a pair of jeans and a ripped red-shirt. He always wears his portable video equipment, which weighs nearly ninety pounds.

Ugly George's raison d'être is to lure women into hallways and convince them to strip for his camera, the love of George, and cable TV. Ugly George is a man who badgers nice middle-class white women until they strip to their soft core.

"Is George Ugly?" wrote D. Keith Mano in a *Playboy* profile. "Does a cat have earlobes? Ugly and getting worse by the half-hour. His long black hair is especially regrettable; old, frizzed, the sort of hair you last saw on a shrunken head. . . . In fact, he may look more like Curious than Ugly George. His skin has the tint of rancid lox . . . and to my knowledge, his long-johns have no understudy.

"And yet," concluded Mano, "this slobbola has picked up more than a thousand women, while you, you're still trying to peek down your secretary's blouse at the office copier."

Ugly George is, essentially, the Zubin Mehta of pick-up sex. He orchestrates. He cajoles. He complains. He kvetches. He has also started his own fan club, calls his life "the story of America," claims to have "an incredibly high IQ," and doesn't quite remember his age.

I love to refer to Lawrence Welk when I talk about my show. He allegedly said that "nobody watches my show except the people." People used to call up the cable systems in New York and say, "Hey, will you hurry up and wire my house? I want to see Ugly George." And the cable company would say, "But, sir, we have films which are critically acclaimed by Pauline Kael." And the guy would say, "Who the hell is Pauline Kael? I want to see some tits with Ugly George! Does Pauline Kael have tits?"

I'll give you another example: Buffalo. I wanted to get my show on a system up there and they sent back my tape with the answer that Buffalo is a Catholic town and "you know the Catholics." A few months later, I did a four-hour call-in radio interview on the phone to Buffalo, and the nerd from the cable company calls up and says: "The reason we can't put you on is that you're not good enough." I said: "Wait a minute! A few months ago, the Catholics in Buffalo wouldn't allow nudity. Now they only allow good nudity?"

What people don't understand is that I educate people. My show has redeeming social value. Lots of people are interested in sex and like to talk about sex, right? Okay. To me, none of that is credible when people have their clothes on. To be nude on a program and to talk about sex is, to me, proof positive, ipso facto, prima facie evidence that what the girl is saying is absolutely true. Also, when she lifts up her blouse, people pay more attention to what she's saying. Don't you see? Like McLuhan said, the medium is the message. What the girls on my show are saying—and what I'm saying—is, "I am here. I'm doing what you would love to be doing. And notice, I am not being paid to do this. I'm a straight, average person—the girl next door." Somewhere, I have a Beetle Bailey cartoon which I love to refer to. The general rushes in and says to Beetle, who's at a printing press, "This newspaper is shocking! It breaks every taboo that exists!" And Beetle looks up at the general and says, "Thanks. But where are all the new taboos coming from?" Well, it's a taboo to be on TV nude but there's no shame in what we're doing. My whole show is anti-guilt. Okay? Why should the word "filth" be associated with sex? That's why I have Freud's picture hanging on my wall. And I constantly refer to him on the show. Fuck guilt! I think that's a tremendous message for our times. It's also one of the reasons why I have engendered worldwide respect, admiration, hate, loathing, fear, and disgust.

People come up to me and hit me. People throw things at me from trucks. Feminists spit on me. I get several death threats a week. And I hear the word "disgusting" at least five or six times a day. Not from old ladies, mind you. Not from the Margaret Dumont types. Never! I always get it from the young pseudo-intellectual types. The liberals. That's why I think there's a tremendous amount of hypocrisy in New York.

The only reason I ever went on cable TV was that every other door was closed to me. They threw me down the steps at NBC. Cable access did give me an opportunity, but's it's always been a means to an end. It's the caboose on the engine of my career. I made it famous and it's treated me like scum.

All I'm saying is that the show has served its purpose on cable TV and has run into a stone wall. I'm tired of Nerd Jerkoff in Jerkwater, Kansas, telling me, "We can't run this filthy show. Oh my God! We have community standards here," and then a hard-R movie comes to town and everyone flocks to see it. Right now, I'm working on my own movie. I've gone as far as I can go on cable TV.

Ugly George held up a recent *Playboy* magazine and pointed to a quote from Brandon Tartikoff, the president of NBC-TV Entertainment: "Ugly George," said Tartikoff, in black and white, "has the most creative show in all of television." He was probably half kidding. In Dubuque, Iowa, as George knows well, the tune was less bouncy. "We have access here, too," said Rod Kuehl, who works for the local cable system. "But you have to remember: this is anti-abortion country and all. We had an access show about a natural childbirth farm, and the woman's exposed breast was a real problem for us. Ugly Geoge? A program like that would blow the roof off this place!"

QUESTION: What do a Rorschach ink blot, cable access, and "The Ugly George Hour of Truth, Sex and Violence" have in common?

ANSWER: When it comes down to it, you are only going to see in them what you want to see.

FOUR

The Local Cable System

In *The Graduate,* a pre-*Tootsie* Dustin Hoffman in the early throes of an identity crisis got a single word of whispered advice on future opportunities. The word was "Plastics." If the movie were made today, there would be two words of advice: "Cable Television." And nowhere on the growing mediascape is the opportunity greater than near your own backyard, in your own hometown. It is there, on a street likely to be called Main or Pine or East 23rd, that the local cable system is situated. According to recent industry figures, there are close to 6,000 cable systems nationwide, and they serve over 30 million subscribers in 13,000 different communities.

The local cable system provides service to the customers along its route and a myriad of job opportunities for the interested or the already involved. Even during the period of the last two years when a critical and prolonged economic downturn forced many industries to reduce or discontinue operations, the cable television industry continued to build new systems, rebuild old systems, and add subscribers at a rate of 275,000 per month. Since 1975, the annual rate of employment in the cable industry has increased over 20 percent each year. Industry insiders are projecting 14,300 miles of new construction and 3 million new basic subscribers in 1983, figures that are expected to result in a significant increase in cable employment. According to Sylvia Marshall, the NCTA's director of human resources, "New Hires in field technical and customer service alone could fill over 16,000 positions."

These cable freshmen (freshpersons?) will be added to a work force that is currently estimated at 65,000—and that's at the system level alone. Many more thousands of people are working in cable television at the program networks and in the supporting services, which abound in cable TV. Others are working at the corporate headquarters of the MSOs—the companies that own the cable systems. According to all the experts and employment figures, however, the real action is at the local system level. At American Television & Communication (ATC), the largest cable-system owner in the country, nearly 95 percent of all the jobs are filled at the local level. And that's a representative figure for all the MSO's.

On the system level, high priority in hiring will likely be given to those areas that generate revenue, such as marketing and sales. Research and development will receive attention as the industry seeks to create and serve new markets. Customer service will increase as the industry strives to meet subscribers needs for service and satisfaction. The 1980s and beyond hold promise for cable as a central part of an evolving video marketplace. Consequently, the demand for more specialized personnel—such as computer programmers—will increase. There will also be a continuing call for skilled and unskilled technicians, from the installer to the engineer. Cable television's manpower needs must be met.

Obviously, not every cable system in the country offers a clean shot at the payroll. In Dangerfield, Texas, for instance, the job seeker won't get much respect. "We have a Ma and Pa operation down here," says system manager W. J. Payt. "We have three cable installers, one technician, and two girls in the office. One of them's my wife." In Dangerfield, the Lone Star Cable Company serves only a few hundred subscribers. Most of them bought cable because there is only one movie theater within a hundred-mile radius.

On the other hand, Warner Amex's Qube system in Pittsburgh,

Pennsylvania, employs 420 (as many employees as Dangerfield has people). More representative perhaps is the cable system in Rockford, Illnois, which employs sixty-five full-time people and eleven on a part-time basis. At the time of writing, there were seven or eight positions available in Rockford, according to the system manager.

Cable television, our newest communications infrastructure, has a growing effect on those who work in and watch it across the country; the industry is alive with varied career opportunities. Some say that cable is now where television was thirty years ago. In other words, whether it be the large system or the middle-sized or the small (but not too small), there is room on the ground floor in cable TV.

THE CABLE SYSTEM JOB FILE

Accountant/Bookkeeper   Many of the middle-sized and larger cable systems hire a full-time accountant/bookkeeper to oversee all budgetary matters. The primary function of this person is to maintain accurate ledgers on both accounts receivable and accounts payable, and to provide a monthly financial statement in preparation for the annual showing of Tax-Man (as in Pac-Man). Annual salary estimate: $15,000–25,000.

Accounts Payable Clerk   A cable system, of course, accrues bills of its own. Another organized soul is the accounts payable clerk, who handles deposits, payment of bills, outstanding purchase orders, and payroll. Often this person assists management and staff in preparation of cost estimates, budgets, and performance reports. Under the direction of the system manager, the accounts payable clerk often functions as an executive secretary, and in the absence of the manager may sign forms or vouchers, approve invoices, and handle other matters according to policy. Annual salary estimate: $15,000–18,000.

Accounts Receivable Clerk   Consumers judge the quality of cable television every single day they watch. But there's one day in particular when they grow most critical and circumspect. It's the day each month when the bill is due and the check should be written. That's the day the subscribers ask themselves, "Am I getting my money's worth? Should I send out the check?" The accounts receivable clerk is the person inside the cable system who has to keep track of the monthly decisions.

The chief responsibility of this person is to keep accurate record of payments by customers. A current list of delinquent accounts must be kept for the accounting department and bookkeeper, as well as the customer service area, to ensure appropriate action. In addition to mailed payments, the receivables clerk must handle over-the-counter payments and may be in charge of petty cash disbursements within the office. There are some people who have a predilection of forgetting where their money is or a tendency to misplace their checks. This is not the job for them. Annual salary estimate: $15,000–20,000.

Advertising Salesperson   Over the past year or two, the local cable system has begun to recognize its potential as an advertising outlet. Satellite-delivered, ad-supported networks such as ESPN, USA, and MTV make time available on their programming so that the affiliate—the cable system—can sell what are classified as "local avails"—advertising spots which can be offered to local businesses. Cable systems now realize that selling "local avails" are a potential profit center for the cable company, and black ink is cable's favorite, though most elusive, color. Many cable systems around the country are now staffing their advertising departments with salespeople who have unswerving will and verve to sell. This is one of cable's major growth areas.

The advantage of cable to the local merchant is price. Cable TV offers exposure in area homes at a cost that is comparable to local radio.

In Wichita, Kansas, for example, a local record shop can buy one hundred thirty-second commercial spots within the MTV channel on the Multimedia cable system in town, and pay just $45 per spot for prime-time (6–10 P.M.) exposure. And the rates go down from there. By comparison, the average thirty-second spot on prime-time network television goes for $150,000. (On the farewell M*A*S*H program, CBS charged—and got—$450,000 per thirty-second commercial spot!) More applicable to the local advertiser are the rates on the local broadcast stations in town. Suffice it to say that none come as cheaply as $45, unless perhaps it's on a Mr. Magoo rerun at 4:30 A.M. on a Sunday. The smart advertising salesperson won't pitch MTV to the neighborhood retirement home, even at $45 a clip. Annual salary estimate: $15,000–25,000 (including commission).

Bench Technician A highly specialized technician, the bench tech operates the cable system's repair facility. Broken or malfunctioning equipment is brought to the repair facility for examination by this person. The technician must diagnose the problem with the broken part, repair it, record the repair, and return the malfunctioning piece to use, if possible. The bench tech is responsible for all repairs, including malfunctioning headends, amplifiers, earth stations, and so on. Why is this technician called a bench tech? Says Bill Sawyers, who performs the job in Greensboro, North Carolina, "Because we work on a bench." Cable television doesn't have to be complicated. Annual salary estimate: $13,000–21,000.

Billing Clerk The primary responsibility here is the computation and distribution of monthly statements. These include the general bill for basic cable service as well as additional charges for pay services (such as HBO, Cinemax, and Showtime). Records, of course, must be kept of delinquent bills and non-pay disconnects, then monitored in conjunction with the accounts receivable department. Annual salary estimate: $13,000–17,000.

Chief Engineer   The top dog of the technical staff, this position requires superior management skill in addition to first-rate technical knowledge. The chief engineer is responsible for all professional and technical concepts of cable system design, equipment planning, optimum layout for cable communications service, specification of standards for equipment and material, construction of facilities, equipment installation, and technical advice to the various staff and operating managers of the system's components. Most chief engineers are responsible for overseeing all the activities of the engineering staff of the cable system. They are instrumental in proposing new services and developing new products for use by the system. If the system is involved in construction—the building of a system from scratch or its expansion throughout the franchise area—the chief engineer will direct this activity as well.

In addition, the chief engineer will participate in market development activities, giving technical advice on franchise acquisition and presenting oral technical updates to municipal representatives (the city council, cable commission, mayor, etc.). The chief engineer also assists in the development of the system's capital budget and general development plans. Annual salary estimate: $25,000–50,000.

Chief Technician   As supervisor of all technicians, the chief tech normally does not work in the field but may be required to do so when complex problems arise (i.e., every screen in town goes blank). The primary function of the chief technician is to assure high-quality signal delivery of satellite and microwave relays to the headend. The headend, remember, is not a piece of cake, from a technical standpoint. And because the equipment is highly sensitive to temperature and humidity, it requires constant control and monitoring. The chief technician is responsible for maintaining this complex equipment. As a member of the supervisory staff, the chief tech must also set performance standards, conduct salary reviews, and handle personnel matters. So not only does he have to know his headend, he has to be a mensch. Annual salary estimate: $20,000–30,000.

Computer Applications Programmer   This position applies only to those cable systems, the handful of them, that offer interactive programming such as the Qube systems offer in Columbus, Cincinnati, and Pittsburgh. The computer applications programmer designs interactive games for the system, helps prepare questions for the interactive polling procedures, knows how to interpret the answers, and understands what it is that computers say when they chat to each other. An understanding of data processing, graphics, creative writing, and analyst skills, not to mention a love for pushing buttons, are part and parcel of this position. As cable matures —and the new technologies evolve—there will be an increasing need for people with that kind of career artillery. For more information, see the profile of Jim George on page 98. Annual salary estimate: $20,000–35,000.

Customer Service Representative   Well-trained "CSRs" are one of the keys to the smooth and successful operation of a cable system. Customer service reps handle customer service calls, orders for installation, and matters pertaining to payments, high expectations, and cable reality.

The CSR acts as an inside sales correspondent, and, as in any system, it's that person who most often comes in contact with the public. The CSR is the cable system's link for informing and selling consumers on cable TV. Many opportunities exist in the customer service area; some people inside the industry, aware of the pressing need for quality candidates, derisively call the current crop of CSRs "the girls in the office." Though it is an entry-level position, the girls in the office will have to mature as quickly as the industry. For more information, see the profiles of Nancy Berg and Liz Serzen on pages 107, 111. Annual salary estimate: $11,000–18,000.

Director of Marketing   It is crucial to the successful operation of a cable system that that system is well marketed to the public. In other words, get the word out, and make each syllable count. The director of marketing develops and coordinates all marketing activity by creating promotional programs, overseeing market research, and supervising advertising to ensure public acceptance of service and products. The director of marketing also must stay abreast of current developments in the field and relay these findings to his/her superiors. In addition, the position has management responsibilities such as interviewing and hiring applicants, conducting salary reviews, and monitoring job performance. Annual salary estimate: $20,000–35,000.

Door-to-Door Sales Representative   The system's sales staff is the "punch" of the marketing program. Some sales representatives conduct house-to-house calls in an effort to educate the community to the benefits of cable, while others inform managers of hotels and apartment complexes. The door-to-door salesperson has to brave rain, sleet, and watchdogs, all in the name of cable TV. This is considered by some an entry-level position, but it is a key position within the cable system. A recent NCTA study found that 49 percent of nonsubscribers indicated that they will subscribe to cable "someday." To the door-to-door rep that day should be today! For more information, see the profile of Frank Freda (p. 102), who has the temerity to knock on doors in New York City. Annual salary estimate: $12,000–20,000 (including commission).

Easement Coordinator   Also called a "right of entry" coordinator, this position is particularly important during the period when the cable system is being constructed and the wiring process is most active. The easement coordinator determines what is necessary from a logistical standpoint to cross someone's property, and is responsible for securing permits to do so, and registering the permits, as well. This person comes into contact with both the municipal officials and the public. Salary estimate: $6–10 per hour.

Engineer   When a community wants a cable system, it asks for bids from parties interested in constructing—and running—the system. Cable companies have established departments to deal with the complexities of the franchising process. An engineer is part of that franchising team. He/She is responsible for conducting extensive topographic studies that determine how difficult the construction phase of the system will be. The engineer also figures out what type of equipment will be needed to build, and where the headend will be placed. There are several different titles for the cable engineer who works in the franchising and construction field, including field engineer, make-ready engineer, design engineer, and construction engineer. Annual salary estimate: $25,000–45,000.

Financial Analyst   A financial analyst is needed during the franchising period, when a programming proposal and construction details are being developed. He/She determines the cost of the proposed plan and also recommends sources of financing for the system. The financial analysis is then presented to the local officials, who will determine who gets the franchise and who doesn't. Annual salary estimate: $20,000–35,000.

Installer   The first rung on the cable technician's career ladder is the position of installer. An installer prepares the customer's home for cable reception, and can be seen in this role atop a sixteen-foot ladder working on the pole, inside one of the cable vans that have become part of America's roadway scenery, or behind a television set inside the subscriber's home, connecting a special terminal device called the convertor. The installer explains the operation of the system to the subscriber and describes the channels and programming the day the service is hooked up.

In addition to the initial installation functions, such as the servicing of feeder lines (the intermediate lines that run from the street to small clusters of homes) and droplines (direct lines to the home), the installer is responsible for the upgrading (addition of services) or downgrading (removal of services) in the subscriber's home. Service disconnects, which occur when the customer no longer wants to subscribe to cable or simply does not want to pay the bill, are a part of the installer's responsibilities. This close contact with the customer requires that an installer know how to smile; the close contact with the sky requires that an installer know how not to fall. In 1982, 85,168 new plant miles were achieved by the cable industry. As more of the major cities get wired and many of the towns and cities expand their cable reach, the next few years will be good ones for installers who can hold onto the pole. For additional information, see the profile of Peter Viik (p. 112), a cable climber in San Francisco. Salary estimate: $4.25–12.00 per hour.

Installations Supervisor   One of those titles that just about says it all. He/She hires the installers, monitors their work, and occasionally handles phone calls from customers who didn't like the installer. A couple more steps up the cable ladder. And throughout the country, the demand for these workers far exceeds the supply. Annual salary: $18,000–25,000.

Market Analyst   Another key job on the franchising team. A market analyst prepares studies to estimate the expected penetration within the community. In other words, how many people will subscribe to the basic cable service, how many to the pay-tiers, and how many won't—and why. Annual salary estimate: $20,000–40,000.

Marketing Researcher   A fact of life in cable television, the researcher is needed at the franchising stage—when he/she prepares demographic studies of the community, which aid in determining the kinds of programs that would best serve the needs of a particular community. For example, if the study shows an active senior citizen population, it will be in the system's best interest to offer a channel that will serve this interest group; if the system is in the middle of the Bible Belt, the National Jewish Network probably shouldn't be pitched. The researcher collects and analyzes all the demographic data to determine what additional programming is needed or which programs should be discontinued. Research studies are also utilized in the expansion of an existing system, determining saturation rate and additional programming needs. For the analytical and detail-oriented, this is an area of growing importance. Annual salary estimate: $20,000–35,000.

Plant Maintenance Technician Sometimes titled pole-line maintenance technician, this classification applies to persons involved in the physical maintenance of a cable plant. They are responsible for tasks that include repairing damaged cable strung between the poles, pole transfer (the actual moving of poles performed in conjunction with the utilities such as the phone and lighting companies), and general repair of damage that might occur to the plant due to adverse weather conditions or other external circumstances, such as a car running down a pole. Annual salary estimate: $15,000–25,000.

Preventive Maintenance Technician This is the person who sweeps the system to ensure that nothing is going sour. One little problem can lead to a big one, and this tech knows a bad slice and a rotten amplifier when he sees one. Annual salary estimate: $15,000–25,000.

Purchasing Agent This position exists at the larger cable systems and at the corporate level of the MSOs (multiple-system operators). Purchasing agents are responsible for buying cable equipment, and choosing from among the many companies that manufacture and sell the hardware. The first step in the purchasing process usually begins with the presentation of products by their suppliers through brochures, catalogues, or visual aids such as slides and short films that show the product being used. Purchasing agents must decide which equipment will be used by the cable system, so they also request references from other cable operators who have used the product, price lists, and any test data the vendor has compiled.

The most beneficial aid are the samples needed for the extensive evaluation process most products will undergo. The key factor, for a purchasing agent, is how the equipment will work in the field. Purchasing agents must stay in strict communication with the manufacturers—and with the changing technology of the cable business. Annual salary estimate: $25,000–55,000.

Quality Control Coordinator. This person works when the system is under construction, and when an outside contractor is brought in to execute the construction work needed. The quality control person does visual inspections of the crew and the work, making sure that the wires are being attached with the right appliances and that safety procedures are being followed. Salary estimate: $6–10 per hour.

Sales Manager Oversees the door-to-door sales personnel, the mailing campaigns, and works with the customer service reps on their approach to presenting the system and its cable services. It is the sales manager's duty to create promotional sales ideas and to implement those ideas. Since there are overlapping responsibilities in every cable system—particularly in the area of customer service and sales—it is the manager who is charged with clearly delineating responsibilities. Annual salary estimate: $15,000–22,000.

Service Dispatcher He/She receives requests for service from the customer services department and dispatches the information to the field. This person monitors service, installation, and trunk-line dispatches, and requests for disconnects, and keeps a record of all communications. The dispatcher may make calls to determine if an service failure has occurred in a single home or is spread over a larger area; this helps to determine the severity of a problem and who should be called in to perform the repair. The dispatcher, after alerting the field, is also responsible for calling the customer's home to verify that someone will be there to let the cable company in, if necessary. The dispatcher's job is a busy one; the cable system in Rockford, Illinois, for instance, receives over 16,000 service calls per year. Annual salary estimate: $11,000–17,000.

Service Technician   This person isn't high tech, but somewhere in the middle. He/She responds to problems from the subscriber, which often require service calls to the home. The service technician will also work on the amplifiers, poles, and lines. This tech is called on to correct electrical malfunctions as opposed to actual physical repair of the plant. He/She must practice preventive maintenance by electronically scanning the system periodically to detect problems before they reach alarming proportions. The person must be well versed in the logic of troubleshooting: working his/her way through a system problem taking signal readings and power readings. The service tech has to be able to trace a problem in the minimum amount of time, and recognize at what point assistance is required. Even the best of technicians know the word "help." Annual salary estimate: $13,000–25,000.

Strand Mapper   This job is vital during the construction phase and when the system is expanding into new ground. The strand mapper will take the drawing of the construction plan and make sure that all the poles are where the drawings indicate they should be. He/She will also walk off the distance between poles so that the construction crews will know how much wire to bring to the job. Another responsibility is to check the position of other wires on the pole: cable wire must be at least a foot and a half away from the utility lines. The strand mapper indicates how the wires are positioned, and how much, if at all, the cable wire has to be moved up or down to avoid tangling with the other lines on the pole. Diagrams must be drawn showing the layout of the system and where amplifiers or line splitters will be placed. Salary estimate: $5–8 per hour.

System Manager   Whether the title is system, operations, or general manager, this person pulls it all together. He/She is responsible for conducting the day-to-day affairs of the system, interpreting and applying the policies of the corporate management (the MSO headquarters), and coordinating all functions of the system. With the help of department managers, the job of system manager includes: recommendation of policies for system growth covering all major areas, including public services, marketing, engineering, and programming; responsibility for managing the system's finances, preparation of the annual budget, and administration of budget and fiscal procedures; development of employment policies, employee benefit plans, and personnel policies that protect the interest of employees and their bosses.

In addition, the system manager will assist in planning system expansion and will evaluate and contract with suppliers for technical material and equipment. In other words, the system manager has to be well versed in every nook and cranny of cable, not to mention every nut and bolt. For further information, see the profile of Ken Knight on page 94. Annual salary estimate: $25,000–75,000.

Trunk Technician   The trunk is considered the spine of the cable system, carrying and making available both signal and power to the distribution sections. The trunk tech must be able swiftly to analyze and react to any trunk system problem. He/She must respond to the loss of power in the system and must acquire an analytical approach to the solution of technical problems.

The trunk tech, like other cable technicians, requires a deep understanding of cable technology and must be familiar with unique test equipment such as the time domain reflectometer and the Wavetek digital sweep. He/She must be familiar with signal to noise theory, signal to hum theory, and the "crossmod" theory of cable television; must have specific knowledge of trunk bridging, distribution, and line extenders; must define loop resistance and perform related calculations; and must have a thorough understanding of the television spectrum, being able to delineate between VHF, midband, superband, and microwave. He/She must also have a comprehensive knowledge of manual amplifier versus AGC amplifiers and the function of which each is capable. To boot, the trunk tech must know how to use an Atlas map. Who said cable TV didn't have to be complicated? Annual salary estimate: $15,-000–30,000.

CABLE SYSTEM PROFILES

Ken Knight
System Manager

In Granite Shoals, Texas, the name Ken Knight means cable television. He is the system manager for the local L-W Cable company, a system that serves just over 4,000 subscribers in Granite Shoals and the neighboring towns of Blue Lakes, Deer Haven, and Marble Falls in Llano County:

Granite Shoals, where the system is located, is about 100 miles from San Antonio, but we might as well be 1,000 miles away. In a small town like this—I mean, Granite Shoals has 635 people in it—no one cares about what Ted Turner is doing, or what the big boys in cable TV are doing in some skyscraper in New York. As far as this area is concerned, I'm the guy they want to complain to. There's only six of us working in cable here and that includes two office people, two installers, a man who is hard to match with a title, and me. And since I'm titled the manager, I'm the guy they want to see.

Which is okay. There are some people I know in this industry who walk around town and say, "Oh, God, I'm in cable TV and I hope no one knows me." You hear a lot of bad things when they know you're in cable. One of our installers here used to wear a cap which said "Damn Cable Man" on the front, because that's what he used to hear a lot. Damn cable man. So there are disadvantages to working in a town that's small. Everyone knows I'm in cable TV, so I have to adjust to it. Inside, I might be saying, "I hope no one knows me" at times, but on the outside I stick out my chest and say, "Hell, I'm with cable TV." It's like a child, this industry. You have to expect the good with the bad.

A typical day for a system manager? There is no such thing. Every day is a different animal. There's always the irate customer and then I act as a buffer. We do our own purchasing of equipment, so I have to make sure the paperwork is processed. I help prepare maps at the office, so that we can begin construction into new areas. That's one of the good things about cable here in Texas. It's open country. If you want to go build in Texas, all you need are the dollars, a permit to go on the poles, and then you build! In most places around the country, you end up getting bogged down with the local governments.

In a small system, you know, a lot of jobs overlap with each other. The other day I was at the bench, repairing some electronics gear; later in the same day, I went to talk to the city council about a rate increase. Or I might be digging a hole in the ground one hour and delivering a franchise payment the next. Right now, we're working on an old building that's been restored. The owners of the building want cable TV so we have to fire the underground.

What does that mean—fire the underground?

Fire means activate. The cable is already in the ground in and around that old building, but it's inactive. So we have to fire it or no one gets to pay for their cable because there won't be any. That's what my job is all about. No matter what it takes, make sure the job is done.

We charge $28.50 a month for the whole package of services.

The basic cable is $10.50 a month and the two pay services, HBO and Cinemax, go for $18. We carry fourteen channels in all, including a station out of Austin and a station out of San Antonio. We also carry the cable networks like USA, ESPN, the Christian Broadcasting Network, and Cable News Network. One of the most common complaints we get is when we switch one channel and replace it with another. People call up and say, "Hey, you took away my favorite channel!" Recently, we dropped a distant network station and put on USA instead. I got a bunch of phone calls from people who said that I took away their soap opera. People here call me at home when they have that sort of crisis, too. I know all their voices, so it's not exactly an anonymous exchange.

It's important, I think, for a system manager to have a feel for his area. One of the problems we have is when we disconnect a house after the person hasn't paid his bills, and then that person tries to get his service back under a different name. In other words, they try to beat us for the payment. Well, around here, you remember the customers. You remember that the same car is in the driveway. In Dallas or Houston or Austin, a customer can get lost in the crowd. Granite Shoals is the type of town, though, where you can find a needle in a haystack.

Another problem we have in a rural area like this is that we can't provide service to everyone. The cost of construction is so prohibitive that we can't build and service an area that won't pay. I'm accountable to the owners of this system. I want to put money in the bank. If there's an area that doesn't have enough houses per mile, that won't return your investment for twenty years or so, then it won't get built. So every day I have to handle phone calls from people who say, "Why can't you serve me?" They think that cable is just a piece of wire tied around a pole that you just hook up to. They say, "Hey, I've got a phone here. How come I can't get you?"

So I have to put it dollars and sense-wise. I tell them how much it costs to lay wire, maintain the plant, and hook up service in areas that have just a handful of homes per mile. Then I tell them what they would have to pay per month to make it worth our while. Something like a thousand bucks a month. Because it costs in reality about a dollar a foot to expand an area. And the reaction I get is, "Hell, I won't pay that!" You have to explain it to them from the pocket, not the heart.

The fact is that people don't *have* to have cable. Cable is like candy. It borders on a luxury. If they can steal it, they'll take it from you. If they steal gas, they'll get zapped. If they steal from the phone company, they'll go to jail. They figure since cable is candy, what can they do to me if I try to steal it? That's one of the biggest problems in cable, and an area which I watch real closely —those people who try to get away without paying for their service. I'm in business to make money and not a killing. We needed a rate increase just to move into the black.

That's a major part of the system manager's job, right? Trying to get the okay from the local city council to raise your rates to subscribers?

We deserve our rate increases. Recently, I went to the city and the city said that I should give the subscribers something in addition to what they already had. That it would help our cause, for the increase, since the people would feel that they were getting something new. So, as I said before, we dropped the distant network station and put in the cable network instead. And we got the rate increase [from $8.50 to $10.50 per month for basic service]. One of the people in the city board, who was serving as mayor pro temp at the time, told us that cable is an optional service. He said that if we need more money to operate, and we ask the consumer for a little more money, let that consumer disconnect if he wants. It's as simple as that. If they don't like it, yank the service. As far as I'm concerned, that's solid thinking. Of course, we do everything we can here to keep them on.

From a career standpoint, wouldn't you rather be a system manager in Houston or Austin? Wouldn't it mean more money for you?

Maybe it's a lack of ambition, but I don't want to go off wandering where the money is. I like to have roots that are tighter. I'm capable of running a system anywhere, and someone's always trying to recruit me, but I'm a country kid. I lived in Houston for a bunch of years and I didn't mind it, but I don't want to fight it again. I mean, I'm a great believer in seminars—the types that they hold for cable people around the country. You're always learning something new in this business. But I only go to the seminars that are held in places I can drive to and drive back from. I'll leave in the morning, stay overnight in a motel, and drive back the next evening.

Where are your roots in cable TV?

I got into it almost by accident because I started working in repair at a research lab. Then I got involved with the building of another system in a neighboring town. Then I became the manager there. You know, I'm not the type of guy with a fancy MBA. I'm a high school graduate and an Aggie; I attended Texas A & M for a while. And I think that when it comes down to it, the most important thing for a system manager is to have a feel for the business he's in. I don't mean that he has to have a college education to prove that he has the feel. If you start out on the entry level in cable, it's gonna get in your blood. You won't want to get out of it. And whether you're in Granite Shoals or Dallas, you have to be on duty twenty-four hours a day. That's what I mean by the feel. I've had the feel for the last twelve years and my wife still gets mad. People on the outside just don't understand.

If you're going to be successful in cable TV, you have to care more for cable than you do for your family. Wait a minute, I'm going to get in trouble for that one. If you're going to make it in cable TV, you have to care for it *as much* as you do for your family.

Jim George
Computer Applications Programmer

A new field in cable television is two-way technology, including the whole specter of shop-at-home, bank-at-home, and interactive entertainment programming. The technology of cable as it relates—

or interfaces—with the computer is one of the emerging career opportunities in the 1980s.

Jim George has seen the future, and he's working at it. At twenty-five years of age, he is a computer applications programmer for the Warner Amex Qube system in Pittsburgh, Pennsylvania, a system that serves over 70,000 subscribers. Pittsburgh is the third stop in the evolution of Qube—and it is an evolution of terminals. The interactive technology of Pittsburgh was turned on in April 1982, and the version offered there is the first eighty-channel cable system. Its home computer console is half the size of its predecessors in Columbus and Cincinnati. It has the capacity to provide any home service, data information, and/or video entertainment programming currently available or, according to a press kit, "that may develop in the future."

Jim George is a Qubeist. He represents the new breed of cable employee. When a computer talks, he listens. He knows the right buttons to push:

About two or three years ago, a writer friend of mine called me up and told me what was coming up in Pittsburgh. I had been working on some talking-head videos, and I had also gotten a computer program analyst degree from the Computer System Institute here in Pittsburgh. It was easy to get into; a lot of these computer schools are called factories, they just pump out thousands of computer people every year. You become what I call a "grunt" programmer, which means you're able to write the computer code, and write programs. I was also going to the University of Pittsburgh at the time my friend told me that cable was coming. I don't know how he heard about it.

After I went to the Computer System Institute, I got a job with a research and development firm that specialized in high-tech training programs. What I really wanted to do was build my analysis skills. Then I heard that this cable system needed a person with analyst skills to work with the interactive technology. After sending a résumé, I remember them calling me at twelve o'clock on a

Tuesday, asking me if I could come in for an interview at three in the afternoon, and by four o'clock, they made me an offer. The next week, I started at Qube in Pittsburgh. I think that tells you a little bit about the demand for qualified people in the computer field. Still, I was lucky. I probably got my résumé in a few hours before everybody on earth—and everybody with a computer degree—started hearing about Pittsburgh, and the opportunity here in cable.

Today, Jim George programs interactive Qube games for the system's Channel 59. There are six interactive programs on the system in Pittsburgh, all done live. "Map at Home" tests the viewer's map skills, using the participatory question—or PQ—format. "Prime Company" matches senior citizens with teenagers in an interactive question program. George works on several of the programs, including "Magic Touch" and "Singles Magazine":

The games are more entertaining now, we're speeding them up. We're into the second generation of games now. At first, we would do simple things pulled out of children's games, or trivia question games, or the game where the first letter of a word appeared, then the second letter, and people at home had to try to pick the word that was coming up. There's no question that the harder the question, the less likely that people are going to respond. "Magic Touch" is second-generation, because it's a game of perception— there'll be a bar scene, or maybe a cook making moussaka. You'll see the ingredients that he's using, and then we'll ask the question: "Which is not an ingredient of moussaka?" The viewer at home then punches in his response. The show runs like a champ. "Singles Magazine" has a host who is a bit comical. He introduces three people to the viewer, and each of the three talks about what it's like to be single. They're seen separately on camera and they don't get to see each other. At the end of the show, we ask the question: "Should Bob go out with Jane or should he go out with Judy?" The viewer punches in the response.

We also have a show called "Qube at Your Service," in which

top management people are put in the hot seat to answer questions or complaints about the cable service. We program PQ questions such as "Do you like this show?" "Do you watch this channel?" The host has to set up the question correctly for PQ to work. Sometimes, the producer of the program will ask me to program a question that is too open-ended. Many producers don't know how it all works. So I'll say that the question isn't right and it won't get a good response. Fact is, you have to understand what an applications programmer is. There are six of us here at Qube and we see over 1,000 questions a day. We work on the show from scratch. And it really comes down to understanding the machine. It's more the machine than television.

What's a day like in the life of a computer applications programmer?

The first thing I do when I get to work is make up my "wish" list—things that I want to do. One of the first things that I get the chance to do is take the morning ratings, to see how many people watched "Magic Touch," for instance, the night before. There's always a little bug in the computer that I have to work on for a bit. Then I work on the games. It takes months to get one together. But that's one of the things I like about the job: I get to create something that was once blank space.

For those interested in pursuing the computer field in relation to a career in cable, how do you suggest they should go about it?

Any interactive programming that we do at Qube is done in "real time." What that means is that things happen right now— and will affect what will happen next. Not everyone does real-time programming, so you need a couple of years working with it. More than anything, you have to learn your machine. Everyone thinks computers can do everything, but you have to be very disciplined with what you want to do with it. You have to remember how quickly things happen on the machine. You have to really develop your machine and analysis skills. I have a very disciplined logic.

The burden is on you to understand what you want from the machine.

It's such a new area, but I would recommend, almost as a necessity, a communications background as it applies to computer science. That means you have to learn how machines talk to each other; you have to understand the "protocol," both standard and nonstandard. And you have to know how to talk back; boom, boom, boom, here's the response.

A data-processing background helps. So does any sort of games background—that is, the structuring of a computer game. If you can make little cartoons on your Intellivision, that's a start. And if you know graphics and have taken a few electrical courses, you might be on your way.

Most university-level computer courses past the Basic and Cobol level, and a minor in computer science, will help, too. I think there's a big explosion coming for those sort of skills in cable TV. With all that background, you can write your own ticket.

Frank Freda
Door-to-Door Sales

The door-to-door salesperson tries to knock down the barriers between the person who won't subscribe and the cable company that says it's the best thing since ice cream on a cone. There are door-to-door salespeople throughout the industry, hired by the cable systems and by independent marketing groups, and then there is Frank Freda. He has been selling door-to-door cable for over a decade, and currently runs a one-man department that he calls the Ministry of Hots for the large Manhattan Cable system in the core of the Major Apple.

"Hot" in cable-chat means a building that has been passed by the wire, and where potential subscribers reside. Freda conducts lobby demonstrations of cable inside the newly cabled buildings, "papers" the mailrooms and lobbies with literature about cable service, and, usually accompanied by the building superintendent or doorman, knocks on apartment doors from the Battery to 86th. When it comes to selling cable television door to door, Frank Freda is the prince of pitch.

I'm coming out with a book called *My Knuckles Are Warm.* I've also written a few plays, which have been produced in my hometown of Philadelphia. In a recent program guide at a small theater in South Philadelphia, I was described as a cross between Harold Pinter and Neil Simon. One of my newest plays, *A Cancer in the Wilderness,* is running now. I should mention that I'm also an actor. I played on Broadway in *Gemini,* in the role of the father, Fran.

I tell you all this because actors make the best salespeople. They don't want to be confined in offices and they know how to press the right buttons. I'm also telling you this because if you know someone who is looking to produce a play, or needs a writer or actor, they can write me. That's why I came to New York in the first place.

Actually, I came to New York to work in cable, while also taking a stab as an actor/writer. I had held a variety of odd jobs around Philadelphia; I had sold some real estate and I also worked at Gulf Oil for a couple of years in inside sales. I didn't have the sanest idea at the time—this was ten, twelve years ago—of what cable was. This was at the time when Newton Minow said that only three people understood cable TV: "One is dead, one is insane, and I refuse to talk about it."

A buddy of mine was probably the fourth person who understood cable. He had been hired in New York to go into the middle of Times Square on top of a flat-top truck and take orders for cable TV. The cable company sent him out not so much to get orders as to have people sign what were tantamount to petitions; they wanted to weigh the figures. The cable company wanted to know: Do we really have a shot here? My friend got a lot of names to sign for cable TV, and he called me and said, "Frank, there's a job here. You should come to New York." I kept telling him no and he kept saying that I should do it. Finally, I said, "Don't ask me again, I'll do it." I figured that my interest was show business, that cable was close, and I knew how to sell.

Going door to door in the mid-seventies was a very confusing period. I didn't really know what it was, and landlords kept asking me, "Are you going to put a nail in my roof?" That's all they thought cable was. A nail in the roof. And the tenants—my potential customers—were even less informed. There were a lot of doubting Thomases around.

What you have to remember is that New York's paranoid about crime. And it really pisses me off sometimes. You knock on someone's door in New York and they won't let you in. They talk to you through the door like you're a murderer. I don't think they would treat murderers any worse than they treat door-to-door people in cable TV. One of my associates had a great idea for getting people to open up. He'd knock on one door on his left, then a few doors down he'd knock on another door to his right, then he'd knock on two doors across the hallway. He'd knock on four doors at the same time. With four different people saying, "Who is it?" one brave person might open the door. Then the rest would figure, "Well, she escaped danger," and they'd open their doors, too. So he'd be speaking to four people in four different apartments about cable at the same time. I thought that was pretty ingenious.

Now, the best thing to do is knock on a door with the super at your side. So they hear his voice, they open up, and I jump them. Once, a woman kept screaming at me through the door to "get the super, get the super!" She knew I wasn't a murderer, but she'd probably just come from a tenants association meeting. They get all pumped up after those meetings. Anyway, I got the superintendent, she opened the door, and when we were alone I told her: "Lady, I just saw your super and you should be more afraid of him than me." I mean, the guy was a caveman. He lived in a grotto.

Personally, I consider myself one of the best ever from a door-to-door standpoint. There are a lot of tired-out satisfied women subscribers who will confirm that, too. (Laughs)

I think my pitch is pretty strong. The first thing you have to be is a chameleon, because you set your style according to the person you're selling. It's not scripted. You have to realize who you're talking to. On the other hand, I have something to counter everything they say. I hear a lot of salespeople around here say, "Well, just start your cable service and then you can drop it." You're not supposed to say that at all. I tell a reluctant customer, "Hey, it costs us over seventy dollars to install cable in your home—and you won't pay the twenty dollars to get it started?" What I make clear to them is: If we don't please you, we lose you; and if we lose you, it costs *us* a lot of money.

I tell my own story, not the company's story. There are a lot of health nuts in New York, you know. All those people who go jogging in back of buses and who will die of exhaust fumes. If I think I've got a health nut—or maybe it's an older person worried about crime—I talk to them about how unhealthy moviehouses are in New York. I tell them that when I was a kid, they would take the side doors down at the local movie theater and the sunlight would sweep through the theater and it would look beautiful. I tell them how clean movie theaters used to be, and they say, "Oh, yes, yes. That's right," and then I say, "Remember: you're a lot safer in your own home from a health standpoint compared to going to the moviehouse of today. Personally, I don't like to see rat droppings on the candy shelf in the theater. I mean, the rats *live* on that candy. And I don't like to pick up a box of SnoCaps and see a couple of roaches scatter."

The wonderful thing about that sales pitch is that when I get through, they're either going to buy cable or vomit. I like the line about the sunlight, by the way. In one of my plays, I wrote that the only time people see sunlight in New York is when they turn around to see if someone's sneaking up behind them, and the sun gets in their eyes.

I guess the easiest person to sell on cable is what I call the over-consumer. He's the type who is totally motivated by something he doesn't have and the fear that someone else has it. He is usually a sexually insecure person, who tries to make up for his unattractiveness with many gold chains and leather pants and cable TV. He's the type who opens his door after I knock and says, "Where have you been? I've been waiting for you!" The toughest sell is the book readers. They are intellectually proud when they say, "I do not have a TV set in my home." Sometimes, I get rather antagonistic with them and I say, "I'm sorry to hear that. Next time a man walks on the moon or gets shot in the street, I'll let you come over to my house and watch it."

One of the areas that I'm heavily into now is lobby demonstrations, in the "hot" buildings that have just been wired. Actually, before the demonstration, I'll paper the mailrooms and the lobbies with literature that tells them about cable. I also paper under doors when I can. I try to get in under there about four or four-thirty, which is about the time people start coming into the house. If you leave the stuff under the door much earlier, somehow it's gonna get thrown out or lost. Once, our literature ran out and I was taken to task by the company for making my own. My feeling was that any advertising is better than none. So I made up my own, and I can bet you that my sales in those three weeks didn't go down.

One of the biggest problems with the lobby demonstration is that people think the TV set you're demonstrating cable on is a special high-powered set. My company believes that if people see how clear the reception is on the set, that's it. In New York City, it's tough to get good reception. Unfortunately, what usually happens as the tenants gather around the set is that one woman who used to have cable in a previous building will yell out, "It never worked right for me!"

Then there are times when the lobby demo can be upstaged. Once I was into my sales pitch and a very beautiful young lady came out of her apartment with her dog to see what was going on. She stayed for a few seconds and then went back inside. She didn't have a stitch of clothing on. Just then, an elderly Jewish woman who was listening to the pitch said, "Nice body. Can I get HBO?"

What is it about your job that you enjoy the most?

I like the detective aspect of it. I like to get to someone's door before anybody else does. It's also a great opportunity to meet charming women. It's not a great job if German shepherds don't like you. The most important thing is your enthusiasm, what you bring to the job. Are you a persuasive person? I remember that when I was a kid, if I liked a movie, then I'd convince all my friends to see it. It's confidence. Some people like to be hollered at. You have to holler at the bosses, too. It's a job for an energetic person. You can't always follow the script that the cable company gives you. I thought I was going to be ill when I first read their sales scripts. You have to create yourself.

I also know that I've had quite a few $1,000-a-week paychecks, including commissions, since I've been doing this. Basically, the salaries run $150 a week base, and you get 30 cents per sale. If you go into a 500-person building and sign up 300 subscribers, it's not too bad. I figure the company owes me a trip to the islands, too. A couple of years ago, they promised a trip to the islands to the top two salespeople in the department. I came in third. Then, when these two guys got back from Club Med, it was discovered that they falsified their records. They lied about their sales. I was actually number one.

So, one of these days, I'll ask for my trip. You see, the difference between me and other salesmen is that "It can't be done" isn't in my vocabulary. That's why, if you're looking for a script or need a writer, I've got the perfect . . .

Nancy Berg
Customer Service Representative (CSR)

None of the high-tech sweat of the technicians nor the high-wire acts of the installers would be worth a drop in the bucket unless the local cable system employed an able administrative staff to handle requests for service, questions pertaining to payments and cable programming, plus the highly sensitive—and potentially voluble—area of customer relations.

The local cable system is a showplace for the service, and also a place to complain. The customer service rep is the person who will try to soothe one disgruntled subscriber who comes into the office to talk about the R-rated flick his eight-year-old daughter saw the night before. The CSR is the person who will try to explain to the subscriber why the movie *Meatballs* has been repeated "ninety-seven times this month." The CSR will be one of those responsible for telling a potential subscriber why cable TV is a much better buy than an Apple computer, a trip to Jamaica, or the Galanos dress spotted on sale.

On the cable team, the customer service rep is the point guard. More often than not, she is the person people will talk to first when their minds turn to cable. In Rockford, Illinois, on the system that serves 45,000 subscribers, Nancy Berg has the calling. She started as a secretary in the summer of 1977, moved over to the switch-board where she routed orders, then became the radio service dispatcher some months later. She is now a customer service rep, and expects to be promoted to supervisor of customer service within a short period of time.

When I started in the summer of seventy-seven, cable TV was going to be a part-time job. A three-month summer job. I was going for my teacher's degree at the time. Then I realized, by watching and listening, how fast people got to move up in their cable careers. It fascinated me.

Here in Rockford, the customer service rep isn't really an entry-level position. There are six of us working full time and two part time, although there are two of us missing now because two of the reps are pregnant. We handle everything from over-the-counter sales to in-office traffic to phone sales and collections. I wasn't fond of the collection duty myself.

According to Frank Sheley, the system manager in Rockford, his customer service rep is more an inside sales correspondent than "a girl in the office." After a sale is made, it is the CSR's responsibility

*to handle the process order, then move on to collection, which is
divided up evenly among the customer service reps. A fourth area of
responsibility includes customer complaints. Says Berg:*

We carry three pay services here, and they give us tips on how
to sell their service. We have a training seminar here that's given
by the marketing people, so we can implement their ideas. Each
CSR is given a selling sheet, which includes the basic selling information, such as it costs $8.45 for basic service, an additional $9 for
HBO, $9.95 for Cinemax, and so on. The installation service is $20.
Once the rep has the basic facts, we're pretty much left to sell
ourselves. We feel out the person, and we should know by the time
they say hello what things we'll be telling. In other words, we'll
know whether to stay completely professional or maybe to joke a
little.

The calls I enjoy the most are the ones from people asking us
about the R-rated movies. We enjoy those a lot. They start out with
the guy joking around and asking how risqué Rockford can get.
That's the type of person you can joke around with a bit. "Shame
on you, sir," I'd say. "However," I'd. add quickly, "HBO has
R-rated movies, and Cinemax sometimes gets a little raunchier."
They ask us about X, but I tell them that we're a conservative town
but we do have R. If a person calls up and is *worried* about R-films,
because of the kids, I tell them that there's an abundance of children's programming on cable that's not available on network TV.
I also mention that each film has several playdates, so that the
parent has the opportunity to watch first and then decide what's
suitable for the children to see. One of the recurring problems I
hear about is that the babysitter comes in and watches an R-movie
in front of the kids. "Well," I say, "we have parental lock units at
no extra charge. This way you can lock up the set and carry the
key in your pocket."

*One of the most common mistakes that customer service reps make,
according to consultant Sheldon Satin, is that they assume the customer knows what cable is. They don't, he reports. A CSR must*

*provide information about the service before they talk about price.
A CSR must sell the features and benefits of cable first. Comments
Berg:*

I think you have to be a person who can handle yourself. You
have to have a sense of excitement about you. The CSR is the
salesperson here, and you have to have experience in dealing with
people one on one. And under pressure. And if you applied here,
you'd have to have some computer CRT [dealing with a keyboard
and a screen] experience. All of our order input goes on the computer now. You don't necessarily need a college degree, but you
have to have good organizational skills.

Do you resent when people in cable knock the CSR?

The stigma here is that we're a group of women who work
together and because of that—the communal feeling—we have it
easy. Some men think we have it made in the shade. Well, it's not
a cush job. It's interesting that we've had a few male CSRs here.
Personally, I'm taken aback a little when a male phone operator
answers. So there's a little stigma there. Plus, the one guy that was
here most recently we tended to mother. I think men are probably
better in collection work, anyway. There's a little more authority
there. Men in particular will assume that if you're a woman, you
don't know anything technical. So a lot of callers think that we
don't know anything.

All I know is that a customer service rep has to stay on her toes.
Once I took a disconnect order from someone who said the customer was deceased. A couple of hours after the installer went out
to disconnect the home, I got a call from the person who screamed,
"I am not dead!" What happened was that it was an elderly couple,
and the husband and wife split up. In order to get back at his wife,
the husband called up and canceled the service, which was under
her name. It was a nasty trick to say that she was deceased. When
we went back to reconnect the cable, she didn't want it anymore.
And when we asked for the HBO box back, which is a $20 refunda-

ble unit, she refused. So now she has an HBO box and no cable TV. But at least she's not dead.

Liz Serzen
Customer Service Representative (CSR)

Liz Serzen, a wife, grandmother, and "an average American who loves the shore," works as a customer service rep for TKR Cable in Warren, New Jersey.

I was working for K-Mart as an area manager in New Jersey, and then we moved to Warren and the traveling between the home and the job became too much. I'd heard that cable was coming to Warren, so I called up and they told me to come down for an interview. Cable wasn't available in Warren yet, so I didn't realize how big it was. When I went for the interview, I told them that I had been a manager for major appliances at K-Mart; you know, things like refrigerators, ovens, and televisions. Their ears perked up. I told them I sold televisions that were cable-ready. They told me they were looking for someone who had sold cable, and I told them that I had sold TV sets. I had sold the idea of cable television as it related to the set. So since I knew what cable was, I got the job.

I'm the first person people see when they come into our offices. I sit in a glass booth across from the front desk. And from inside that booth, I try to hook the fish. And I have the bait. I tell people that if you go to the movies, you have to park your car, and pay four bucks a person to get in. Popcorn is $1.50. If you went to see *E.T.* with your entire family, that's over ten bucks for the night. I tell them that if they buy Spotlight, it costs only ten bucks and they get sixty movies a month. So that's fifty-nine movies for nothing! That's a hook. *Mommie Dearest* on cable was a hook. *Camelot* on HBO was a hook.

My job is to sell cable TV. My job is to make sure that check is coming in. Then I have to get the checks to the bank and make sure they're hooked up. What with hundreds of checks a week and hundreds of hook-ups, my day goes like a shot. I make sure they're happy, too. Sometimes they'll call me up and ask me about WHT [the STV service offered in the same community]. I tell them that they get one channel over there for $35 a month. End of competition. For a $20.95 installation fee, and $25.50 a month for HBO and Spotlight and all the basic services, I think cable's a real buy. Just send that check for $46.45 and your account is set up. No hidden charges. No contracts. That's my line—you're not contractually bound. They have to feel like they're not trapped.

I get paid a commission rate, and though I make under $20,000 a year, it's going to improve, I think. It's pretty good for a woman the way it is now. After the corporate level rolls the dice, I'll see how they settle. But I hope they still have a job for me as they expand. Sometimes they tell me, "Be patient." Hey, it's a learning process for me, too. But I know, if it came down to it, that I could peddle elsewhere.

Peter Viik
Installer

Cable installers know all there is to know about upward mobility. They have to climb poles for a living. They are also the customer's most visible contact in cable TV. When it comes to hooking up the service, an installer can't come on like Attila the Hun.

The installer position is considered an entry-level slot in the fast-growing field of cable TV. However, both cable systems and home owners pay very close attention to their work. Cable installers lay it on the line; a mistake here or there could result in an unhappy, *ex*-subscriber. When that happens, it can get very lonely at the top of the ladder.

Peter Viik is a cable installer for the Viacom System in San Francisco, which serves over 70,000 subscribers. He believes that installers have to be "psychologists in hard hats":

We go into people's homes to hook up the cable service, and it gets sticky at times. You always have to be sure that you're not stepping on the customer's toes. We come in with wires and tools and some people think you're going to rip apart their house. "How big is the hole that's going to show? How much wire is going to show? Watch out for that lamp!" We get those sort of questions and comments all the time.

The best thing to do is to go through the basement with the wire. This way you have access to route it throughout the house. Everything is run from the ceiling of the basement. Then you drill a hole in the floor above and put the wire through it. The better the wire is hid, the happier the customer. Once, though, I drilled a hole in a floor, went down to the basement, and couldn't find the hole. Another time I drilled right into a water drain, and we had to replace it. Most of the time, though, it goes very well. Most of the time, installers don't blow it.

San Francisco is built really strange, of course, so using a ladder here proves to be very entertaining. There are houses on top of houses and all sorts of hills. Sometimes, you have to climb over a building to get to a customer's house. Once, I was climbing over a roof and I heard an argument going on in the house above me. All of a sudden, I saw a woman throw a man out the window, and he came crashing down on the roof. I went over to see if he was all right, and he told me he was having a little trouble with his reception.

Things get a little wild on what we call "non-pay" days, too. That's the day once a month when the cable installers go out and try to collect delinquent accounts. Once, I was chased up a pole by a guy with a gun. "I paid my bill," he screamed. "Don't cut me off!" Things get pretty hairy. That's why installers don't wear three-piece suits. You have to be able to run from the crazy, irate customers.

The biggest danger, though, is climbing the pole. The cable goes right from the pole to the person's home. And when the weather is bad, when it's really windy as San Francisco can get, it's hazardous. A few days ago, twelve of our poles were knocked down by the wind. And when you're up there, you'll dealing with thousands of volts of power. On days like that, the company will call us in. They're very safety-conscious. They hold safety meetings every week. . . .

And don't forget about the dogs. The other day, I went to a customer's house, started walking up the stairs to the TV, and was confronted by two huge Doberman puppies. They leaped on me and I almost had a heart attack.

Why in the world would anyone want to be a cable installer?

It's a very pretty thing. Sometimes, I climb up a pole, look out, and see the San Francisco Bay. You can really appreciate your job when you see a sight like that. I'm from Brooklyn, New York; if you climbed a pole there, there'd be nothing to see. Here, there's a nice air about it. I can smell the coffee roasting in the morning.

You work as a cable installer because you like to smell coffee?

There are other reasons. One of them is job security. It's a union position here now, and there is a tremendous need in the industry for people who know how to do what I do. Another is the sense of responsibility. After you've gone through the training period, the whole thing is yours. You carry around several thousands of dollars worth of equipment. You get to drive a truck without some guy looking over your shoulder. It's your baby, and you don't have anyone bugging you. I don't want to get stuck behind a desk. Installing cable is a good clean honest day's work. And because you have that independence, it sort of weeds out the good installers from the bad ones.

One of the biggest problems in cable TV is the piracy of service. Can an installer be bribed by a customer to hook up service without notifying the cable company?

It is a problem. An installer can be faced with temptations, such as a person offering him money to keep his service on even though the customer hasn't paid his bill. Personally, I think more of the future than a quick bill stuffed in my pocket. There's too much happening in cable TV for me to do crap like that.

How else does the problem manifest itself in the field?

Some people go out and buy coaxial cable from Radio Shack and splitters and try to hook up the cable service themselves. The problem is that it's illegal, and the cable often isn't shielded right, so it not only causes problems on the person's set but feeds right back into the system. When the wire is not shielded properly, the signal is allowed to radiate; and if there's enough of those wires, the signals end up going into the air. There have been reports around the country of pilots coming in for emergency landings, looking at their instruments, and all of a sudden getting Showtime on the screen. Radiation could interfere with the landing of planes. It's a real problem. But since our installers are eventually going to be in and around every home in San Francisco, we're trained to detect the piracy of cable service. Personally, I don't know any installers who would accept a few bucks and risk their job in the process. There's something about staying on the job, too, that tends to weed out the good from the guilty.

You've been on the job since when?

I first started at Viacom in 1972. I really was fascinated when a guy came in to hook up my service for the first time. It was a weird thing. The guy was bursting with energy and he was originating something. He was very excited about his job. I was the one with all the questions. About a week later I called up the company to ask about the job and a guy named Bill Womack kept telling me, "No openings, no openings, no openings." He ran everything out of a small room. But I was very determined. It had an aura about it. I *had* to get involved.

After a couple of weeks of hanging around the office, waiting for a break, I was finally told to go grab some tools. I was a stranger in a strange land but I learned very quickly. The hands-on training was great.

I've traveled a good deal of the past few years, too. I've worked at different Viacom systems across the country. New York. Florida. My brother, who is six years older, tells me that I've wasted all this time—that I should've got a degree. I tell him that the degree I got is what I learned about people. I got to see the U.S.A. And I enjoy my work. I still get psyched up when I take out the truck in the morning.

FIVE

The Made-for-Cable Networks

The made-for-cable networks have taken the sameness out of the television screen—which was made to be stared at. Made-for-cable networks were made to be paid for, and every time a subscriber does so, cable TV is called a horn of plenty. When the subscriber decides to spend his cash on his kid's teeth instead, that's when the made-for-cable network has to shake the spit from the horn. How about the twenty-four-hour-a-day Orthodontist Channel?

The amazing thing is that there are currently over forty national made-for-cable networks, dishing up the kind of programming that can turn on, in spots, an audience that is all-black, all-adolescent, all-healthy, or all-American but proud to be horny. Expectations aren't the only things that cable has raised. But beyond the made-for-cable moans, there are cable networks that can educate (those are the ones that are lauded and hardly watched), and cable networks that can kill you with gloss. One of the wonderful things about having cable is that you can keep totally up to date with twenty-four-hour news and completely out of touch with the street. When you have cable television—and you blow the horn of plenty —you don't have time to wonder what else is doing besides TV.

The simple fact is that the made-for-cable network, starting with Home Box Office in 1975, has changed the face of the industry. Its promise of diversity has given cable the cover of *Time* and the sort of headiness that comes with having an ace in the hole. Everybody just knew that network TV was for the bleary-eyed birds. The feeling was that the only people who enjoyed network TV were the people who shopped at K Mart. Cable television winked, and planted the idea that it understood. And it sold the idea that movies, lots of them, on TV were better even than Mork and Mindy. Then came the concept of cable networks *without* the movies. The idea here was that if you had heard of tofu and wanted to die old, how could you resist Cable Health?

The national made-for-cable networks can be loosely grouped as those for which subscribers pay a separate fee—the "pay" or "premium" channels like HBO, Showtime, and the Movie Channel— and those for which they do not—"basic" channels such as USA, ESPN, and Ted Turner's Cable News Never Ends. The "basic" channels, of which there are forty-five, make their nut by selling ad space, or at least trying to. Some of them charge the cable system for the privilege. The "pay" channels earn their keep by sharing the subscriber fee with the cable system operator who offers the service. The term "pay" channel is a bit confusing, because cable subscribers pay in some way for anything they get, at the very least by making monthly payments for their cable hookups. The term "basic" channel is a little confusing, too. How can anything called MTV—and which offers video clips of Def Leppard—be termed "basic"?

There is no confusion over the fact that the made-for-cable networks have provided jobs, in addition to a considerable load of programs for the viewer who likes to switch. HBO employs 1,200 people, and the National Jewish Network 3; ESPN counts 350 heads, and the "pay" channel Spotlight runs a payroll of 45. The Playboy Channel, which jiggles, pays 60.

CBS Cable and The Entertainment Channel employ none, since both cable networks fell into the black hole with losses large enough to make a grown financial analyst sob. There is a virtual guarantee that at least a few more made-for-cable networks will flutter and fail, too. It's called shakeout time in cable land.

But the ones that survive—and there will be
tinue to need quality folk. There are plans for n
too, which means more jobs on the horizon
ongoing demand for programs—and the people who ...
marketeers, who can sell cable networks like they love their moth-
ers; financial types and their No. 2 pencils; and a host of others who
can help make a cable network make its space.

At last count, there were 25,000 people working with the made-
for-cable networks. If you're looking to break in, made-for-cable
means you've got a twenty-four-hour-a-day chance.

THE CABLE NETWORK JOB FILE

Advertising Sales Like Richard Burton and Elizabeth Taylor,
advertisers and cable television have had a stormy affair. The
"basic" channels such as ESPN, Cable News Network, and ARTS
—to name three of thirty such services—need to attract advertising
dollars to stay afloat in the sea of plenty. An uphill struggle might
be the best way to describe the current state of affairs. Many
advertisers are still reluctant about throwing dollars the cable way,
citing insufficient "numbers"—such as the Nielsen ratings—and
misdirected programming concepts.

Nevertheless, many of the major advertiser-supported cable networks *have* been able to land significant sponsor accounts, and it
is the responsibility of the advertising salesperson to scour the land
in search of the advertising green. The advertising salesperson must
work with the marketing department of the particular network,
and have a thorough knowledge of the numbers available, in addition to all the reasons why his/her network is the best thing to
happen to cable TV since John Walson's storefront window. A
good deal of phone work—and foot work—is required in this
position. An advertising salesperson for a middle-sized cable network reported that she had met with eighteen different agencies,
and thirty-four different account executives, in a period of three

weeks. At a rate like that, the pitch doesn't have a chance to get cobwebbed. Annual salary estimate: $15,000–30,000 (plus commissions).

Affiliate Relations Representative   The bottom line for every cable network is to be seen on as many cable systems around the country as possible. The cable system carrying the particular network is called, as in broadcast television and radio, the affiliate. Home Box Office, for instance, has close to 4,000 affiliates across America; the premium cultural network entitled Bravo has 40. It is the affiliate relations representative who is on the road, meeting with affiliates and making sure that all is well, and also meeting with cable systems that do *not* carry his/her particular network.

Many of the cable networks have regional offices, where three or four representatives in each office are responsible for cable systems in that area. Some of the smaller networks have only three or four affiliate relations reps, and they are responsible for the entire country. The affiliate relations representative offers marketing tips to each individual system carrying the cable network, so that the system can better serve its customers in relation to that particular program service. The representative must also be available to soothe and woo any system manager or program director who has a question about the cable network, or a problem with it.

The representative is the voice of the cable network in the field. When it comes to meeting with cable systems that are reluctant to carry the network because of channel space restraints or other reasons, the affiliate relations rep has to make sales calls, which include slide presentations and examples of that network's programming. Many reps like to say that they have to "enlighten" the nonaffiliate and "educate" the system that is an affiliate. The representative often spends time with the local system's customer service rep (CSR), who is the person the subscriber, or potential subscriber, speaks to when there is a question or complaint. It is important for the CSR to know what to say about a network, and the affiliate relations rep from the cable network has to be good at putting words in other people's mouths. Annual salary estimate: $20,000–35,000 (plus commissions). For more information, see the discussion with Jane Taylor on page 159.

Assignment Editor   Again a position found more frequently at cable networks that rely on immediacy: the news channels and the almost-all-sports channel. (ESPN recently added a financial report in the early morning.) The assignment editor decides which stories need coverage and which are best left to the *National Star.* He/She makes sure that the network has the story blanketed, in terms of on-the-spot reporters and camera people, and also negotiates with the stringers in cases when they are needed. The assignment editor must be very organized *and* creative in the ways of getting a story. He/She must also know how to rip copy from the telex, which keeps a steady stream of late-breakers and updates flowing into the newsroom. Annual salary estimate: $30,000–40,000.

Business Affairs   Since it is called show business and not show art, there is a need on the network level for a person (usually a lawyer) to review negotiations and contracts with anyone the cable network does business with: agents, producers, writers, directors, unions and guilds, distribution companies, and so on. Once a network programmer, for instance, makes a verbal commitment to proceed with a project, a business affairs person is brought in and consulted. Often, the business affairs representative will negotiate for the cable network and try to secure the best possible deal for the network, given the financial and legal circumstances that the programmer has presented. Business affairs people vary in color and height, in sex and weight, but they all operate under the same credo: there's no hurry. Since it's their job to find the pony in the manure—from a corporate standpoint—they have to be painstakingly *thorough,* although the people on the other side of their desk often use descriptive adjectives of a slightly different nature. Annual salary estimate: $35,000 on up.

Creative Services   This heading covers several different areas of responsibility that have varying titles, the most common of which is communications coordinator. Creative service people provide the artwork needed for the network's program guide, in addition to the layout and structuring of the guide. Many cable networks set up booths at the major industry conventions, which include promotional material ranging from high-gloss programming booklets to gifts bearing the network logo and name. The creative service person is responsible for the design of the booth and the materials, in addition to the monthly affiliate kits that are sent out from network headquarters to the cable systems around the country. A creative service staff at the cable network level normally includes writers, who are responsible for the words that go into the guide and the other network promotional packages. Annual salary estimate: $15,000–30,000.

Director   In cable television, a director is responsible for translating the script—or the concept of the producer and network—into a form of entertainment. The cable director is no different from any director in that regard. The difference between a cable television director and a network television director is the money he/she has to work with (in cable, it's less) and the types of shows he/she has to direct. One of cable television's bread-and-butter forms is the "On Location," which HBO originated and the rest have followed: the taping of a full-length concert or performance by a known star. The On-Location shows are also called in-concert shows, and they fall under the heading of Variety TV. Last time anyone checked, variety was dead on network television. On cable television, it's alive and well, although the directors are running out of stars to shoot.

A director in cable TV might also be called upon to direct his cameras at a stage; plays-for-cable television are enjoying current popularity. One of the newest areas for directors are the "made-for-cable" movies, that is, a movie financed by a cable company and designed to air on the cable network before it is seen anywhere else.

In all forms of cable TV programming, the director is responsible for rehearsing the actors, assuring proper lighting, and getting in the "can" the movement and magic that a television show can be. In post-production, the director often is responsible for editing the pieces into a whole and ensuring the proper flow and rhythm. The director in cable TV is responsible for the look of the show; through his vision an idea can be guided from paper to performance to the point when people are paying to watch it on cable TV. (For more, see the profile of cable's premier director Marty Callner on page 141.) Annual salary estimate: varies.

Director of Communications   A managerial position, this person heads up the creative service department and is ultimately responsible for the affiliate material and program guide, and often acts as a liaison between the marketing, public relations, and creative service departments. Annual salary estimate: $25,000–40,000.

Driver-Messenger   A classic entry-level position for college graduates and aspiring cablephiles who want to get a toe in the door. The driver-messenger will shuttle video material to and from airports in a company car, and distribute mail through various departments on foot. He/She will also shovel the walk if the snow has blocked the entrance to the office. In other words, a man or woman for all seasons—and any job that needs to be done. "It's a great exercise in humility," says Barry Black, the human resource director at ESPN. "There's nothing but the wall behind you in the pecking order. But you get to meet valuable contacts. And it gives us a chance to see how you work—for when a promotion comes up." Salary estimate: $4–5 per hour.

Field Engineer   Also called "operational engineer," this high-tech troubleshooter makes sure that the distribution of the cable network service—from uplink to satellite to cable system—is in good working order. If there is a problem with transmission, the field engineer must know how to solve it as quickly as possible. If a cable system is experiencing recurring difficulties with the network signal, a field engineer travels to the system and trys to pinpoint the trouble. Annual salary estimate: $35,000–50,000.

Financial Analyst   Unlike the program estimator, who ascertains the feasibility of a single project from a financial standpoint, the financial analyst judges financial matters with a long-term perspective, not project to project. This person looks at trends in the cable industry, often projecting a network's position over a five-year period. The financial analyst works with department heads in terms of budget, and coordinates closely with the corporate heads of the cable network in order to achieve a clearer crystal ball. The analyst also will be in on the ground floor of major corporate decisions, such as the moves into pay-per-view showings, and keeps close tabs on viewership patterns, subscriber numbers, and advertising revenues.

Some cable networks call this person "vice president of research"; no matter what the term, this position requires a thorough knowledge of cablenomics, and the ability to interpret changes in the cable industry. The financial analyst often writes up major reports and studies, which are read Bible-like by the network bigwigs and whoever else can get their hands on them. Since the reports usually contain numbers, graphs, and arrows, they are stamped confidential. Annual salary estimate: $40,000 on up.

Human Resources   Some cable networks call this position "personnel," even though human resources sounds a bit more impressive. The human resources person is responsible for hiring above the entry level, works in career planning and counseling for the network employees, provides personnel evaluations for the network brass, and serves as a liaison between top management and lower-level employees. The human resources professional (or personnel director) will also get involved in any management training programs the network might offer, the intern programs, and in shaping company policy concerning employee incentives, working conditions, and so on. Annual salary estimate: $30,000–65,000.

Intern   At least 20 percent of the existing cable networks offer some kind of intern program, which enables a student or cable novice to work in a variety of roles within the network organization. Usually, the internship is a nonpaying position, but the payoff often is a job with the company. For more information on intern programs, see Chapter 7, An Education in Cable TV.

Marketing Service Coordinator   There are currently several outside companies—supporting services—that publish weekly and monthly cable guides for the consumer at home. The marketing service coordinator acts as a link between the cable network and the publishing entities, ensuring that the network schedule is included and that program highlights are given appropriate notice in the publications. It is important for the marketing service coordinator to establish a good working relationship with the editors of these cable magazines. Annual salary estimate: $17,500–30,000.

Newswriter   At cable networks such as ESPN, Cable News Network, and Satellite News, the newswriter supplies the words that the on-camera anchors read. A newswriter is responsible for writing succinct, well-paced news copy almost as quickly as the news happens. Usually, a newswriter will get the basic information from the wire services, such as AP and UPI, and then add quotes and material as these flow into the office. The newswriter has to be a clear thinker and a good organizer; in most cases, he/she will also be called on to learn how to use a word processor, which helps ensure the rapid flow of copy from writer to producer to the newsreader waiting for something to say. Annual salary estimate: $12,000–20,000.

On-Air Promotion   This network job entails working with an independent company, or in house, on the on-air promos that announce upcoming events such as a special concert production, a blockbuster movie, or a soon-to-be-seen series airing on the network. If these promos (usually short taped messages designed to capture the viewer's attention and imagination) are to be produced by an outside company, the on-air promotions person must work with that company to ensure that the "look" befits the network, and that the company has all the information it needs to manufacture the video clips. If the promos are done in house, the on-air promotions person could be responsible for producing the segments. Annual salary estimate: $25,000–40,000.

Producer Most of the cable networks buy their programming from independent producers, while the basic cable webs such as ESPN, Cable News Network, and Satellite News Channel employ staff producers who are responsible for the many hours of live programming that go out on those networks. In the case of the independent producer who is not employed directly by the cable network, but instead sells programs to HBO, Showtime, and other cable networks, it is this person's responsibility to initiate and create projects (i.e., by hiring writers to flesh out original ideas), to seek out properties to be adapted to cable TV, to package projects (by finding and attracting writers, directors, and stars), and to prepare budgets for the shows. The producer is also responsible for preparing the treatment, which describes in print the basic concept of the proposed program(s), and a rundown (a running time flow of the proposed show and its elements). Finally, the producer must set the meetings with appropriate buyers at the cable networks: for example, the head of comedy development at Home Box Office for a comedy series proposal.

Producers are considered the prime movers and shakers of the software business; they must develop projects that are right for cable television and they must be able to "package" them in such a way that they will be both creatively and financially sound. It is not enough just to have a "good idea"; at the cable network level, the package—the collection of names attached to the project—is considered at the very least on a par with the idea itself. The producer must not only develop the concept of a show but have good connections to the creative community. The top producers are usually the top salesmen, too. Since it is necessary to raise money in order to manufacture a show, the producer must be able to represent his show well in meetings with cable network programmers. The cable network, after all, must write out the check in order for the cameras to roll.

It is the responsibility of the producer, then, to stay on top of all trends in the cable business, both in the United States and in Canada, and hopefully to initiate a few trends on his or her own. Once the show has been financed by the cable network, it is the producer's job to supervise its execution by keeping track of the budget, representing his/her independent company in dealings with the unions, supervising the production managers, and making sure that all goes according to schedule—usually the network's schedule. When the program is completed, it is the producer's job to bring it over to the cable network, for a screening, while thinking of his next project on the way. (For more information, see the profile of Chuck Braverman on page 132.) Annual salary estimate: varies greatly.

Production Coordinator   In the cable network hierarchy, the production coordinator hardly rates a nod. It is considered just an inch or two above entry level, and the salary is commensurate with the lack of clout. Still, it's a job that many college graduates would die for. The production coordinator will do everything from feeding the staff to ripping copy, to communicating the need for new typewriter ribbons. The production coordinator, by the nature of the job's position on the floor, will also be privy to all the insanity that takes place at a network and to the people who have the power to promote. It's a wonderful spot to be in if you want to listen, learn, and begin the climb up the ladder. Annual salary estimate: $9,000–15,000. For more information, see the profile of production coordinator Carrie Petrucci on page 152.

Program Estimator   The person at the cable network responsible for breaking down the script with regard to cost, or deciding whether a program venture is worth it from the financial perspective. The program estimator is utilized in the development stage of a project prior to any commitment by the program executive. Annual salary estimate: $30,000–40,000.

Programmers   Since it is the network's aim to present programming that is either markedly different from the competition (rare) or just different enough to be called "about the same as the rest" (not so rare), the programming people are responsible for creating, acquiring, shaping, and finally airing shows—or "software"—that will best represent their network from a financial and philosophical point of view. The cable network programming executive is the one who will decide, flanked by committee usually, which program ideas to develop and buy and which ones to turn down—or "pass on." The programmer must read the treatments, peruse the names attached to the shows, look at the budgets, circulate them around the office, and then, if the caller is lucky or has enough clout, take a phone call to discuss the project.

The programmers at the two major premium cable networks, Home Box Office and Showtime, report that they are pitched thousands of program ideas a year, through the mail, on the street, by phone. Since they are "buyers," they're often called arrogant or imperious when in fact they are merely swamped, Unless you walk into their office with Marlon Brando signed, Robert Redford in tow, and Steven Spielberg next to your shoulder, the odds are that the programmer will take a few days, if not weeks and months, to get back to you about your idea.

There are several different types of program executives; your idea will be considered by one—or several—of the following:

Head of Programming   In some cases, the president of the cable network serves as the head of programming, or at least the person who shapes the programming philosophy of said network. He/She is responsible for the final decision, particularly on projects that will cost a great deal. He/She is also responsible for the entire on-air image of the network or, as some in the cable industry like to say, the "perceived value" of the network. For more information, see the interview with Michael Fuchs of HBO on page 164. Annual salary estimate: $75,000 on up.

Program Development Person The larger cable networks have program development people specifically for comedy programs, documentaries, made-for-pay movies, drama series, and sports programming. The program development person is responsible for sifting out the good from the bad—at least from the network standpoint—and for meeting with producers and other suppliers in connection with potential projects for the network. Program developers will read scripts, treatments, letters—and between-the-lines —to discern whether a project is worth mentioning to a vice president of programming or the head of the network. Often, a program developer will offer suggestions to a program supplier, and provide insight into what the network is looking for in terms of the particular concept.

One of the worst moves a producer can make is to ask a program development person: "What are you looking for?" If the idea that's been presented is close enough, he/she will probably provide some input along those lines. If you're a country mile apart, you probably won't have more than a quick handshake goodbye. Annual salary estimate: $40,000–75,000.

Interstitial Programmer This is the person who fills the space between the movies and specials and the on-air promos and regularly scheduled series. Also called "shorts programmer," he/she is charged with finding quality programs that run for only two to ten minutes, usually not much longer. Traditionally known as filler material, the shorts can help give a program service a unique look and appeal. The interstitial programmer doesn't develop these programs, but has to view the hundreds of tapes sent in from around the land. The cable networks don't pay a lot for interstitial programming, but the job is important, nevertheless. A good eye—and good contacts in the world of filmmaking—are pluses for the interstitial programmer. Annual salary estimate: $20,000–40,000.

Publicity A good network publicist gets the word out about the programming, people, and principles that make the network special, at least in the mind of the company president. He/She will deal with editors and reporters from the cable trade papers (see Supporting Services), set up press interviews with corporate bigwigs, write releases that are sent to newspapers and magazines, and arrange for clips of network shows to be used for publicity purposes. The publicist works with the marketing department at the network to put into effect promotional gimmicks and PR largesse, which all help to increase the "perceived value" of the particular network. The publicist is also helpful in squashing rumors, letting the cat out of the bag to the right dogs, and keeping a smile on the corporate face of cable TV. Annual salary estimate: $30,000–60,000.

Scheduler This network position could be described as the glue of the finished product: the programming that a network puts on the air. The cable network scheduler is responsible for positioning each program in such a way that the viewer has the best opportunity to see the shows and the shows themselves have the best chance of finding their audience. Given the diversity of typical programming fare, the scheduler must be a master juggler and have a good sense of the potential viewers for and value of each particular show. He/She must know when to put on the show at 8:00 P.M. and another at midnight. For more information, see the profile of Alan Zapiken of Showtime on page 155. Annual salary estimate: $25,000–40,000.

Stringer   The news-gathering forces in cable TV, CNN and Satellite News, and the virtually all-sports channel ESPN, often hire freelance on-camera reporters and camera people to cover late-breaking stories in areas where the networks do not maintain bureaus or staff personnel. At ESPN, which covers sports 23 hours a day (there's a business report program in the early morn), more than 150 stringers—or "freelance" correspondents—contribute video reports to the network, and the turnover is reportedly quite high.

Many of the stringers across the country work for local television stations in their markets, but others work strictly freelance. In Portland, Oregon, for instance, a reporter/cameraman team turns in over fifteen different stories a month for Cable News Network, for which they are paid approximately $400 per news story. Since there is a continuing need for stringers in some of the lesser markets, reporters and camera people are advised to send their sample tapes and résumés to the news producers at the respective cable networks. Salary estimate: $300–500 per taped report or story.

Tape Librarian   Since all the cable networks have to deal with an abundance of videotapes culled from the network schedule, the tape librarian is needed to handle the filing of tapes in such a way as to make them readily accessible for future use and research purposes. When Bear Bryant, the legendary college football coach, died, the ESPN producers needed a clip from his final game, which they found in the tape library in the network office. The tape librarian must devise an efficient filing system so that the needed tapes are quickly found; the librarian might also be responsible for disposing of tapes at his/her discretion if the network has space restraints. Annual salary estimate: $15,000–25,000.

Traffic Coordinator   This person works closely with the scheduler, and is responsible for putting together logs that indicate the flow of material to be used between the motion pictures that a pay

service airs or the features that a basic cable network has scheduled. The traffic coordinator makes sure that the flow—or "traffic"
—of product is documented, and that all programming is ac-
counted for. Annual salary estimate: $20,000–25,000.

Travel Coordinator   This position is more prevalent at the basic
cable networks that rely on live programming, such as CNN and
ESPN. The travel coordinator is charged with arranging all travel
plans for staff reporters, producers, and camera persons who have
to hit the road on assignment. This includes taking care of plane
tickets, hotel reservations, auto transport to and from the airport,
and sometimes the distribution of "per diem"—the small sum of
money given per day to traveling staff people for food, taxis, and
other amenities. Annual salary estimate: $15,000–25,000.

CABLE NETWORK PROFILES

Chuck Braverman
Producer

Charles Braverman graduated from the University of Southern
California Film School in 1969, in a class that included George
Lucas. After a brief stint in Air Force Reserve, Braverman went
to work at CBS News as "the best messenger they ever had." He
left to work on "The Smothers Brothers" staff for a year and during
this time produced a three-minute film called *American Time Capsule,* which documented—in a fleet, neat way—the history of the
country. He formed his own production company after "The
Smothers Brothers" was canceled and began to make a mark as a
producer/director of commercials, short films using a process
called kinestasis (which he had utilized in the *Time Capsule* piece),
and documentaries.

In 1978, Braverman produced his first cable program for "Showtime": a program featuring highlights from the Oscar-nominated

short films of that year. Since then, Braverman has produced many original programs for Showtime, including the "What's Up, America!" series and the "Big Laff-Off" series featuring young comedians in stand-up competition. With five years of cable credits, Chuck Braverman is considered one of the producing pioneers. He worked in cable television before other producers had heard of it:

In the beginning, cable was a lot more free than it is now. You could do things by the seat of your pants. In network television, you have programming and practice people, who are responsible for watching every word you tape, and you have network executives, and then you have someone called a vice president of something who wants to see the dailies [the daily taped material] and your first cut. But in cable, they'd say, "Okay, you want to do a show about Willie Nelson? Fine. Go do it." And that was before Willie Nelson was America.

Several years ago, I suggested to Showtime a ninety-minute music special with Willie Nelson, and they weren't sure who Willie Nelson was. Then his picture appeared on the cover of *Newsweek* and they wanted to do it. So I was able to produce a full ninety-minute Willie Nelson concert without having guest stars in it like the Rockettes or Carol Burnett. That couldn't be done on network television. They would have wanted a bunch of guest stars. But on cable TV, you can present a purer form of entertainment.

There used to be a stigma about working in cable. When I would call agents bout talent that I wanted, they wouldn't know what cable TV was, so I would have to explain it to them. Now, everyone's looking to get in. It's like a new candy store has opened and everyone wants to get their fingers sticky. It's the new media. For producers, it's another place to sell product; to sell shows. It's expanded the market. Everyone wants a piece of the action.

From a producer's standpoint, has cable television changed in the last five years? Has the market changed?
I think Showtime and even HBO have been innovative in their

programming. I think they have not been innovative *enough*. The problem now is that cable television is getting more and more like network television because a lot of the people who are working at cable networks are from NBC, CBS, and ABC. Cable TV has evolved into huge, moneymaking corporations run by three-piece suits. They're less willing to take risks because they have to answer to big corporation people. I don't think they have any real adventurous spirit in the professional filmmaker's sense. You know who their audience really is, who they program for? The guys who own the small cable system in each of the local towns. That's who the networks are programming for—the system owners have the real power. They're not even programming for the *people* who are watching in those small towns; it's for the guy who owns the cable system. If that guy likes the program, everything's fine. And it's getting so that he's an older, wealthier, more conservative kind of guy. He's the guy who decides whether Showtime is going to be on his system. So you don't want to offend him.

There's a pendulum effect in cable programming. It swings back and forth, and right now it's swinging toward the conservative side. I just had a piece pulled from my magazine show "What's Up, America!" which wasn't exploitative, it wasn't dirty. A year ago, they were pushing us to do more "R-rated" pieces in every show. Then they told me that some of their Southern affiliates—the system owners again—were complaining. So Showtime told me, "We're pulling this piece," and that's the first time it's happened.

What was the piece that they pulled out of the show?

It was about a casting agency for pornographic films. It was a very very good piece: an inside look at the porno business without the exploitation. There was some nudity, but nothing overt or outrageous. What was outrageous in it was how brutally honest everybody was in the piece. And I think it was somewhat startling to these three-piece suit, corporate types, who are sitting back in their offices watching the piece, afraid of offending the Southern system owners, or the Moral Majority, or somebody.

Still, there's more freedom in cable than on the network level. You just have to know what sells. Most of the people that I know who are in this business don't have any idea of what's commercial, of what the cable networks want. That goes for producers in cable, too. Some of them have good ideas for pieces of shows, but they don't know what's on the air—they don't know what the buyer is buying.

How do you find out what the buyer is buying?

You read. Everything. Like I do. You read the trade papers and anything else you can get your hands on. I probably subscribe to more magazines and newspapers than anybody. It's my wife's biggest complaint at night: "It's between me or the newspapers." But I know what's commercial.

As an independent producer selling to the cable television market, what steps would you recommend for other would-be producers?

Stay out of the business. There are enough of us already. Don't even buy this book. Go back into the shoe business. Go to work for your uncle.

You don't want any more competition?

I've got enough competition. (Laughs) I don't want to encourage anyone to be knocking on Showtime's door. Don't even tell them the address. Don't tell them the phone number!

You see, it's cutthroat out there. If you're on the outside, they probably won't even take your call or read your treatment. Fortunately, I'm not a kid in my twenties any more. I'm a kid in my thirties. I have people working for me. I've become part of the establishment, I hope.

Then you can afford to give advice . . .

I think the best thing to do if you want to be a producer in cable is to try to make an appointment with me, or one of the other major suppliers. I'd apply at every decent production company in town and I'd say, "I'll do anything for any amount of money." Don't say, "I'll work for free." No one should have to do that. But say, "I'll be your driver or janitor or gopher." That's how you get your foot in the door. The guy who's the director of cinematography for me started as my driver. One of the guys who's considered one of the top feature film editors in town started as my driver. My drivers quit and become producers. You can't come to me as a film school graduate and say, "I'm a great cinematographer. Trust me." Because I'm not going to trust you. Neither is anybody else.

You become a messenger or mailboy first. You'll find out who everybody is. You'll find out how much everybody else is making. You'll find out who's doing what to whom, and where. And then you move on. If you become the best messenger at CBS, like I was at one time, then you become valuable enough so that when you're ready to move on, they'll say okay, because they know you—they know you're going to be good.

The thing is, I was a messenger at CBS News and I had my college degree from the USC Film School. And I learned a hell of a lot. I learned how the inside of CBS News works, which has helped me as a producer. I assimilated things that I didn't even know I was learning.

The point is you're not going to be able to walk in to HBO or Showtime with just a good idea and your charm. Those people aren't going to give you $200,000 or $300,000 to do a show. But if I walk in with Joe Blow from Podunk who has a great idea, they know *I* can deliver the show and that *I* will be financially responsible. They know they'll get the show on time. The same thing goes for me, by the way. If I want to make a major motion picture for Warner Brothers for $20 million, they're not going to let me do it. But if I walk in the door with Dan Melnick or Ray Stark or David Putnam [well-known movie producers], it gives me credibility.

How does an unknown would-be producer with a good idea get in touch with Braverman?

I think it's important how the person approaches me, or any of the other independent production companies. If they have a great idea, they should try to get an agent. I'll always take a call from an agent because I know he knows the business. Or the guy with the idea could write me a note or call me. Now, I'm probably not going to take his call cold, either, because I'd spend all day doing that. But the guy could drop me a note without giving away his idea. And then I might be willing to talk to him about it.

Why did you hire the last person that you hired?

It was a mistake. It was a first impression, a gut reaction. And it was a mistake. Usually, my first impressions are right with men and wrong with women.

It's funny, 'cause I want someone whose appearance is nice. It's true. Maybe that sounds a bit chauvinistic. If it's a man, I want him to shave every day. I'm sounding like an older, conservative person now, but I think it's important. I want somebody who is bright and uses his head and is aggressive for the company. Someone who's always thinking, who has some street smarts. I've had some secretaries; there was one lady in particular who could speak a half-dozen languages fluently, who could type 100 words per minute, and who was an absolute genius. But she had no street smarts—she'd let anybody in off the street, and if you smiled at her, she'd let you take the company typewriter with you.

I want to continue producing cable television programs. I've made more than my living in pay cable over the past five years. But I need the type of people around me who won't let other people walk off with my typewriter!

Bill Tush
Cable Network Superstar

Since 1974, Bill Tush (as in gush) has served as Ted Turner's court jester and cable television's original all-purpose funnyman/anchorman/star. He started with Turner before the superstation was super, when it was just good ol' Channel 17 in the broken-down house on West Peachtree. When the station and Turner's cleft went national—via the SATCOM satellite in 1976—Bill Tush came along for the ride. Over the past nine years, he has done on-air promos, anchored a ridiculously late evening news show, hosted a comedy show entitled "Tush," a series called "The Lighter Side," and currently presides over "The Tush Show," the talk program that is sent out from Los Angeles. Tush replaced Mike Douglas as the host when the show was called "People Now!," and is seen rubbing ad-libs with major stars, and other odd types, who come on the panel.

Hollywood—and his current state in cable—are a long way from the days when Tush shared his anchor's desk in Atlanta with Rex, the Wonder Dog, a German shepherd who was fed peanut butter right before air time so his jaws would move as if speaking. Off camera, Tush would do the doggie's voice. When Turner got serious about news and propped up CNN, the edict from the top came, as usual, in memo form. Wrote Turner: "Tush, no more talking dogs!"

We actually pulled off the Rex bit quite well. That was in the days when our attitude was "No news is good news." So we thought we'd do it funny. Once, I read an entire news show while holding a photograph of Walter Cronkite in front of my face. Another time, I wore a gorilla outfit while reading the story of a guerrilla attack. Then there was the time we had someone scream off camera while I was doing the news, and I got up, ripped off my shirt to reveal my Superman outfit, and jumped over the desk to save the lady in distress. I occasionally let loose with a flying cream pie, too.

I cringe when I watch some of the old tapes today. I also remember wearing a W. C. Fields mask during a newscast, and then getting a phone call from my parents, who said: "Why don't you do the news seriously?" It was tough to explain why I would do the news some nights with a bag over my head. That's when I was doing the "Unknown Newsman" bit.

Before I came to Channel 17, I was in radio in Atlanta, at

WGST. Then in the spring or summer of 1974, I began watching the station and they were running a lot of old movies. I know old movies, so I went by to see if they needed a booth announcer, someone to introduce the films or just do some promos. I came in with a demo tape that I had made, and I asked the receptionist who I should see about the job. She was a temp and didn't know. Then an engineer came by and told me that their staff announcer had just quit the day before. So he listened to my tape, and he told me that if I did one hour of promos for the station, they'd pay me fifty bucks. Well, I got my fifty bucks for the hour, and then I began putting in an extra nine hours here and ten hours there. I was only getting paid for one hour, but I was so happy, I wouldn't leave!

Then I finally got on the air: someone put a slide up and I read the news over it. My hair was shoulder length, I had glasses on, I didn't own a suit, and I borrowed a weird-looking jacket and tie. It was horrible, but no one complained. And if Turner's happy, he's happy. It's when he starts rantin' and ravin' that you know you're in trouble.

When I was finally hired as a full-time on-camera newsman, I got $12,000 a year. Up the street at NBC, the weekend *weatherman* was making $25,000. But things started to happen around here, and when we went on the satellite, I knew that I'd outgrown the local stations. I enjoy being a national figure. That's what keeps me here. The place is hopping. It's not a cracker-box local station.

I was in New York recently and someone yelled at me in Penn Station: "Hey, I watch you on the superstation!" Then I was in San Francisco, walking into a men's room, and some guy comes out just as I'm going in and says, "Wait a minute, aren't you that guy on the Atlanta cable station?" I said that I was, but he didn't believe me. He kept saying, 'Oh, c'mon, are you *really* him?' So I showed him my driver's license.

I'm a performer. I love the attention. And I'm not at the point where I'm mobbed by two hundred people. Every now and then, someone nods in my direction. That's damn exciting for a regular guy like me.

Has your fame increased since you moved to Los Angeles?

A friend of mine does a very good business in bus-shelter advertising. You know, the ads that you see at the bus stations? He put a big ad at a bus stop at Sunset and Vine: *Who is Bill Tush?* The funny thing is that I went from being a very well known person in Atlanta to a nobody in L.A. Tony Curtis comes on my show and I can see him thinking: "Who the hell is the guy with the name of Tush?" However, there have been several advantages in coming out here. The contacts that I've made are incredible, that's number one. Also, I'm no longer in awe of meeting famous people. When stars used to come into Atlanta, they'd treat you like you were a local yokel. Now that I have my own show in L.A., I'm on a more equal footing.

When I came out to Los Angeles from Atlanta, the company got me a bigger car, too. I went from a Chevy Citation to a Buick Regal. I have this great fear that I'll pull up to a car wash in Beverly Hills and the maitre d' will say: "Sorry, we don't wash Buicks here."

If it all ended tomorrow—or if Turner sold his networks and got out of cable—what would be the legacy of Bill Tush?

When I get that bus ticket to has-been city, I can always go back to Atlanta or some small-time city station. Or, for that matter, I had lunch with Turner the other day and told him: "I'll tell you what—just let me feed the bears on your plantation." He said that if worse came to worse, I could always feed the goldfish in front of the building. I don't see that happening, though. The secret to my success is that I've just crept up to where I am. And I'll keep creeping.

You didn't quite answer the question. What of the Tush legacy? How will they write of you in the history book that gives a chapter to cable?

When the history of cable is written, I'll be mentioned in association with Rex, the Wonder Dog. That darn dog will follow me to my grave. And the thing is, Rex isn't even in cable any more. I think he's doing the news for a TV station in Chicago!

Marty Callner
Director

If there were a single man responsible for the Home Box Office "look"—a look that can glitter and gloss—then a strong case could be made on behalf of a man named Marty Callner. Of course, there are several people responsible for the look that has hooked more than 13 million homes to date, but Callner certainly has made a serious contribution to the cause. He is, simply put, pay television's most prolific director and, arguably, among the top two or three directors working in television today. Callner operates with an exclusive agreement to direct for HBO in the pay-TV field, and his eye and cameras have snared one hundred thirty specials for the cable giant, dating back to 1975. From HBO's standpoint, Callner's eye is as big as Cyclop's. He has directed—he has *shaped*—many of HBO's most memorable moments, including the landmark George Carlin show (and the seven dirty words you *can* say on cable), the Steve Martin special, and Robin Williams' first step outside the eggshell called Mork. He has directed many of the best HBO music shows, including those with Diana Ross, Pat Benatar, Fleetwood Mac, and Hall and Oates. And, to his ever-lasting credit, he lived through the HBO presentation of *Camelot,* which he directed as both a stage play and a pay-TV production. *Camelot* ended up a beautiful show, but the scenery backstage took a beating.

At this point in his cable career, Callner is more than a director: he is a cottage industry. But although his fame and fee have grown since his first special with Robert Klein, Callner continues to take chances in an industry that can be disturbingly status quo. A Marty Callner show on HBO looks expensive but can involve viewers instead of shutting them off. His shows may glitter, but they're rarely glitzy. He understands that even music shows can have drama.

Even before he had a beard, which works very well on directors and psychotherapists, Callner was a star director or, as he puts it, "the leader of the laundromat." He started as a prop man for WCPO-TV in Cincinnati, Ohio, where he grew up, and was promoted to staff director in seven weeks. Who can forget Rosemary's brother in "The Nick Clooney Show"? Callner moved on to Cleveland and directed numerous national commercials. The next stop on the map was WBZ in Boston, where he directed the Celtics basketball games, Red Sox affairs, and just about anything else that moved. On the wings of a recommendation from then-Celtics announcer Dick Stockton, Callner began doing some freelance work for a new and unheralded cable television venture called HBO. The year was 1972, and his fee was $200 a game.

I liked the years when we were the bastard child of network TV. No one knew what HBO was, and there was pure resentment against us. When I left WBZ in Boston, I had the chance to go to NBC Sports or to HBO, and I chose HBO for a lot less money. Intuition is the name of my game. When I was sixteen, about twenty years ago, I bought a thousand shares of Cox Cable. The shares were going for $3 each, and I told my mother to buy cable. I had a strange feeling about it. A strange psychic feeling. So we bought a thousand shares, and we sold them about four years ago at $66 each. We made a fortune. I had the same feeling about the Playboy Channel last year. So I bought ten thousand shares. I trust my intuition.

The funny thing was that when I turned down NBC and took the work at HBO for about $50,000 less, people thought I was crazy. The network people used to put me down, and they used to put HBO down. I remember that we did a special with Liza Minnelli once, and the next day ABC's "20–20" was going to shoot a feature on her. Michael Fuchs [HBO programming chief] and I told them that we'd leave our lights up so they could use them for their show. They looked at us and said, "Cable TV lights? We don't want to use your cable TV lights!"

The next day, we went into the hall and one of the ABC people said that they'd use our lights after all. They knew they were the best lights in the business. Michael and I told them to put the lights where the moon don't shine. We ordered every single one of the lights down. We were ready for a fight. We were ready to fight all the people who put us down.

Of course, now all those same people are scared. Scared to death of HBO. David has become Goliath and there's a great deal of self-satisfaction. And as far as I'm concerned, I love to be loyal to cable TV. It's far better than the networks in every respect. It's the coolest, most creatively free way of doing television. It's encouraged me to stretch. Network TV, you know, is tied up by the same five or six guys in town. Cable television has a very large talent pool, a creative force that previously couldn't find work. Pay-TV was created from the street, by a bunch of entrepreneurs and gamblers. There's still a rough edge to us. And that means it's just starting to be mined. It's a gold mine, that's what it is. All you need is a big enough pair of balls to get in on the fun.

And it would help to have the eye of Marty Callner. How would you describe the Callner style of directing cable TV?

Each show that I do is the best show it can possibly be. Each is right for what it is. I'm a director first, then a television director. And my imprint is that there is no imprint. It's seamless. Invisible. People aren't supposed to notice the shots. I never repeat shots. I don't want to become bigger than the artist. I see it happen all the time, even on HBO. A director doesn't trust his material. Other directors try to dazzle you with their effects. My shows are beautiful because I'm less obtrusive, and I believe that lighting is next to God. I'll never allow the artist to look bad.

My biggest success is that I've done 130 on-location shows and they all look different. I do interpretative directing. That's my strength. And I leave nothing to chance. One of the ways I make women look beautiful is that I always have a fan blowing on them, to keep the sweat off their face. It's a variable-speed fan that I control from the director's truck. I shoot them through filters, at the right angles. I always explain to them what is happening. I teach them what TV can do. It's a magnifying glass. If someone is really good, it'll make them seem great. If they're the least bit bad, it's going to make them seem worse. That's TV. It puts you under its thumb. So you have to prepare the artist; you have to make sure they understand how Machiavellian the TV camera can be.

I'll give you a metaphor. I used to raise and train horses in Kentucky; I went to school to learn how to do it. You know, the biggest goal of a trainer is that on Kentucky Derby day, his horse is the best that it can be. Maybe not the best horse in the field, but the best that it can possibly be that day. The horse is geared from the time it's born, from the food it eats, to reach its maximum potential of speed on Kentucky Derby day. I want my artist to reach maximum potential on the day I shoot. It's a motivational way of directing.

I knew that Pat Benatar was going to be great the first night that I shot her; we always tape two consecutive nights, and the second night is usually the one we use for the show. But, from talking with Pat, I knew that she's the type of woman who shoots her wad and then it's over. She doesn't do well under pressure. So I brought her along—and my staff—to be ready for that first night's performance. I knew Diana Ross was going to be better the second night. She was a little weak the first night, and I brought her to the second show by letting her know that it was okay, that we were a little weak, too. I made her feel like we were on the same schedule. And when we are—and we usually are—it can be magic. We can touch people with our shows. Most of my shows are made with love.

What about Camelot, *which was perhaps your biggest assignment for HBO and one of the industry's most important shows? It certainly was the industry's most expensive show at the time it was done.*

Camelot *was made with hate. There was a tremendous amount of dissension between Richard Harris [the star] and the producer,* and I sided with Harris. There was a tremendous amount of back-biting within the industry, too. All the cable people who hate HBO —and there are a lot of them—started calling it "Spend-a-Lot." It is the director's responsibility to shape the program and to see it through, and my grit got us through *Camelot*. It was a spectacularly beautiful show. And it gave me a lot of credibility. I've gotten a number of movie offers since that production. But I also realized that no show is worth that much pain. It meant that I didn't want to be famous any more.

John Huston once said, the higher you go up the tree, the more your ass shows. People think of me as a creative genius who sits in an editing room by himself. I'm not an esoteric, Woody Allen–type person. But I do have a great fear of failure. I believe in the power of karma. And I'm also more me in my work than I am anywhere else. My defenses are down when you look at my work. And my work is not cheap. If I've done anything for cable television, I've made sure that the productions aren't cheap. My original budget on the Diana Ross special was $250,000, for instance, and it ended up at $750,000. I spent close to $200,000 on the set and lighting alone. All I said to HBO was: "Listen, this is what I'm trying to do." And they said: "Okay, let's go for it. But it better be good, mother."

That's what I mean about cable TV. They're not afraid. And you can't be afraid, either. If you want to be a director in cable, you have to be existential—you have to be responsible for what you do, and you have to live your dreams. I have a favorite saying: "This is no dress rehearsal." Cable TV is no dress rehearsal.

I started in the industry at two hundred bucks a shot. I wasn't opportunistic, I just had a feeling. Directors have got to have instincts. If you do something just for the money, you'll never make any. And you have to believe in the medium in which you work. If you don't believe in cable, it will have nothing to do with you.

Janice Thomas
Advertising Sales

Janice Thomas is the director of advertising sales for the Black
Entertainment Television (BET) network, a basic cable service that
offers a mix of black-oriented programming to close to 4 million
subscribers nationwide. It is Janice Thomas's job in advertising
sales to solicit potential sponsors for the network and to ensure that
present-day advertisers are satisfied with the placement of their
commercial spots. Ms. Thomas has a masters degree in communications from Howard University, and began her television career
at WRCT, the NBC-owned station in Washington, D.C. Cable
television—and the network BET—presented for her an opportunity to advance more quickly up the employment ladder and to
meet bigger challenges:

At the NBC station, I was the sales/research coordinator, and
I analyzed the Nielsen and Arbitron audience statistics. I would
pass along the data to the sales department, which was in charge
of getting sponsors. I had been there for about a year and a half
and I wanted to do more. And I wasn't satisfied with the pay. NBC
is such a big corporation that I didn't see myself getting promoted
and I didn't see much upward mobility for the future. I didn't want
a career at NBC. So I started entertaining the thought of redoing
my résumé and looking for another job. My experience in looking
for jobs was that it took six months or more to find one. So I figured
I would be at NBC for a little while longer.

About this time, I got a call from a lady named Charla Beales,
who worked at the cable association office in Washington. I had
met her a couple of times, and she called to tell me that Bob
Johnson, the president of BET, was looking for someone in research. So I was lucky to know someone who knew the head of a
cable network. What I didn't know was much about cable TV. All
I knew about cable, really, was that I had watched it when I'd go
to New Jersey for summer vacations.

I'm still learning about cable TV, too. I like the challenge of learning. And I also like the challenge of what I'm doing—selling cable to potential sponsors. Or trying to. I knew it would be a difficult process, but I have to be resourceful. I'm running the advertising department and doing what five different departments do at NBC, so every day is a new hurdle that has to be hurdled. That's the incentive about working here and in cable: we always have to push harder.

One of your main areas of responsibility is to fly from the Washington office to Madison Avenue and try to woo sponsors for BET. Is that as difficult a process as, say, selling shaving cream to Arafat?

There are some people, you know, who have no interest in cable at all. They tell us there's not enough organized research about it, and when there isn't enough research, they don't want to hear from you. Sometimes I'm told: When you have the numbers, call us back. But not before you do. Our feeling is that sooner or later these people are going to have to get involved in cable TV, so we send them information about the network. I usually know pretty much by the greeting I get at their office how they're going to respond. When you've been at the job for a while, you begin to size up attitudes very well. They key thing is not to get frustrated. Some advertisers still treat cable like the plague, but you can't let it get to you. You have to give them as much information as you can and then, on the way down in the elevator, think, "There's always a next time. And the next time, they might very well buy."

How do you try to sell advertisers on a cable network that appeals to the black audience?

Black people, traditionally, have been ignored by advertising as well as by the three networks. You don't see many black people on television shows and commercials. Yet, as a population group, black people watch more television than any other group. That's a fact. They actually watch twenty to twenty-five hours more per week. Black people also buy the products that are seen on TV. There are many many products that the black person buys.

Our cable network is providing a forum for advertisers interested in reaching that important group of people. The black family enjoys our black-oriented programming, and we sell that fact to the sponsors. More often than not, sponsors tell us that it's a great idea. But the same sponsors might tell us that they've run out of money in their budget, or they say they'll get back to us. Sometimes I hear: "We want to see how it grows." When a sponsor doesn't go with us, I think it's because there's a resistance to cable, not the specific concept of BET.

Do you see the situation improving over the next couple of years?

My feeling is that it will become an easier process over the next year or two. Cable—and BET—is growing very fast. It's here to stay. And the more experience we have as an industry, the better I'll be at my job.

In what ways will you be better?

I'll become better at what I do because there will be a need for me to get better. You know, you have to sell yourself as much as the product. I would like to get more aggressive and I think I will, as I get more comfortable in the job. One of these days, I think I'll be running an advertising department with several people. That's a challenge I have to meet head-on. The bigger cable gets, the bigger the challenge. It's wonderful. That's the reason why I left NBC in the first place.

You know what you have to think when you're selling cable TV to potential sponsors: I can do it! That's what you have to believe. No one ever said the job would be easy, you know.

Martha Quinn
Star

Prior to landing her job as a prime-time VJ (video jockey) on MTV, Martha Quinn was best known for her role as the Chicken McNuggets girl on a TV commercial which aired only in New York. Now, at age twenty-four, she is known as Martha Quinn to millions of cable subs who tune in and turn on to MTV, the phenomenally successful channel that serves up video music clips supplied by record companies. interviews with rock stars who can speak, and concert films, all culled from the rock 'n' roll experience. MTV is an FM radio station you can watch. With her engaging manner and impish grin, Martha Quinn could make MTV fans out of people who swear by Mantovani.

My career plans didn't include cable TV at all. I majored in broadcast journalism at college, and I had taken a course called "The Advent of Cable," or something like that. I'm sure it was a required course, because all my energy was focused into getting a job in radio. That's where I wanted to be. I mean, I went to a radio station that had a college. I spent one year at Colgate University, in upstate New York, and while I was there I did a radio show every Tuesday morning. For three hours, I played a lot of Steely Dan and James Taylor and The Beatles, and I basically ignored the traffic and weather. It was a free-form ten-watt station and I had no idea what I was doing. I was sort of like, "Okay, it's Tuesday morning, I think, and here's James Taylor." I loved it. I played what I liked to listen to. And even though I got into it by accident, it still had enough glamor for me.

How did you end up doing the college radio show by accident?
   The guy who was doing the morning show got sick one day, and he lived on the floor below me in the dorm. That day, I was passing through the hall and I heard: "Hey, can someone do my show? I'm dying down here." So I decided to give it a shot.
   I've always wanted to do something public, I guess. I was in all my high school plays and I wanted to be an actress at one point. I got through college doing TV commercials here and there. I admit that I was the girl who said: "You'll go nuggets for McNuggets!" I'm not proud. The first week I was at MTV, someone

stopped me and said, "Wait a minute. Aren't you the McNuggets girl?" I think I told him that I had just moved to America from Afghanistan. What's a McNugget?

I should tell you that I transferred from Colgate to New York University in Manhattan. I wanted to be in radio, so I thought it would be a good move to come to a city where the stations have more than ten watts. I think you should go to college near where you want to work, so you can grab an internship and maybe work your way from there. You have to put yourself in the place and position to get lucky. The good thing about cable is that it's strong in Cleveland and Columbus and Atlanta and Houston, and in a lot of other places where you can't find a job in radio or regular TV. There are a lot of centers, big and small, in cable TV. But since I wanted to go into radio, I thought I'd make my big connection at WNBC.

What did you do at WNBC in New York?

I got Tabs for everyone, unless someone wanted a sandwich. My big project was sitting alone in a closet and putting the singles in alphabetical order. After a while, I got very resentful. I can make programming decisions, I thought. I can shape the radio industry! I was fetching lunch. I told my professor about it, and he said I should stick it out. I was a real good student and I did. Well, my internship ended and I ended up getting a job as a receptionist at the dormitory where I lived. There were no job opening at WNBC. Then one day, my professor said that if I wanted to, I could go to work for a station that he owned, out in Riverhead, Long Island. It wasn't a real big station and it was only a weekend job—I knew it would cost me more money to get out there than I would make —but it was radio. I thought it was a fantastic start, and a way to make it back to New York.

So where did MTV fit in?

A few days before I was going to start my weekend job at WRIV, I was in New York doing some shopping, and I thought it would be nice to go up to WNBC and see a few of the people I'd become friends with. I wanted to tell them that I got a real radio job. So I went up, went around, asked people to stay in touch.

I was in the middle of saying a few quick goodbyes, when I overheard a guy asking about Bob Pittman, who had been program director of WNBC a couple of years before. "What's he doing now?" Someone from across the room said he was involved with MTV. I wasn't in the middle of the conversation, by the way. I was just catching little bits and phrases. Out of the corner of my ear, I heard the guy say, "I'm going to call Bob right now," and out of the corner of my eye, I saw him pick up the phone.

Out of the blue, a friend of mine from the station comes up to me and says, "Hey, Martha, you should be a VJ!" I asked him what a VJ was. He told me and I said, "You mean someone who sits there and spins records on camera?" I had no idea what he was talking about! All of a sudden, he gets on the phone with Bob Pittman, who was talking to the other guy, and says, "Are you still hiring VJs?" He hands the phone back, turns to me, and says, "They want to see you in twenty minutes." Twenty minutes! For a job interview! And I was wearing shorts!

When I went into the MTV office, I didn't even know what to say to the receptionist. "I'm here for an audition or something." So I was brought into a room, read some music news off a Tele-PrompTer (an electronic cue-card machine), and then they said, "Tell us about a concert you've been to." I'd been to about four Earth, Wind and Fire concerts and I rapped about it for a couple of minutes. Someone said, "That's enough," and that was it. A five-minute audition.

Two days later, on Thursday, I stopped by my apartment to pick up a picture and résumé, which I was going to bring by the MTV office, when I switched on my answering machine and the message said, "Call us. We have some good news for you." It was a fantastic feeling. I got the job! It was a break out of nowhere! I was really excited when they asked to speak with my lawyer about the contract. I mean, I would have signed on a paper towel!

Interviews with Yoko Ono and Elton John. Personal appearances. Eight-by-ten glossies that you have to sign. Do you ever lie awake at night and think: "God, what would I be doing today if I hadn't gone shopping that day?"

I really think that I was very, very lucky. What if I hadn't stopped off at WNBC? What would have happened if that man hadn't called Bob Pittman? It was an amazing set of circumstances. But I'll tell you: we have about ten interns working now at MTV. One of these days, one of them might get as lucky as me!

Carrie Petrucci
Production Coordinator

Can a non-salaried intern find happiness and a paycheck in cable TV? Carrie Petrucci's story is emblematic of the industry. She worked as an intern for Cable News Network during her last quarter at UCLA Film School, and used that experience to land a job at Financial News Network, a basic cable service available in almost eight million cable homes. At twenty-three, she's a production coordinator—which means she does a little bit of everything. Though low on the totem pole at the network, Carrie Petrucci is receiving a paycheck, and has nowhere to move but up. Now that she's on the inside, she's in the position to do so.

I guess the most important thing is to keep a third eye out, to see how you're appearing to other people. If you appear hardworking, and not out only for yourself, you can move up. You can't appear overly ambitious. In college, I always used to hear: "Be nice to everybody once you get a job. Don't make enemies." I guess you can't go up to someone and say, "It's stupid the way things are being done." You have to be very discreet. But on the other hand, you can't be a brown-noser. And you can't lose your own values. I mean, some people will do anything to move up or get the job in the first place. I'm not sure about that. I guess you can move up if the right person likes you. Right now, all I want is to have a job where I can use my brain. But I guess you have to do the shit work to get to do what you want to do.

I get up every morning about four-thirty, because I have to be in the office by five-thirty. My hours are five-thirty to two in the afternoon. Right now, we're on the air for seven hours every day, although there are plans for us to expand to twelve. My most important responsibility can be the most boring. Every half hour, I keep a running list of the most active stocks on the New York Exchange. It's supposed to be an indicator of where the stock market is going. So I speak to the news bureau at the Stock Exchange, and I get the information and we relay it on the air. I also rip scripts, starting at about seven every morning. The writers in the newsroom type on six different carbons, and I put them in order and distribute the copies to different groups in the office. We go out live, so it's important to get the scripts to the anchorperson on time. During the course of the day, I'll also fill in for people who go to lunch. I also do the graphics operating. I did a lot of typesetting work to pay my way through college, and I can type real fast, so I started to learn about the graphics machine about fifteen minutes every day, until I edged my way in to working it. During the course of a typical day, I'll also keep the Dow Jones numbers straight and handy, I'll fix the typewriters, and I'll get food into the newsroom, for people who can't get away to eat.

Sometimes I think to myself: "I went to UCLA Film School for this?" I mean, it's not intellectually satisfying at times. So I come up with projects and ideas. I read the business section of the newspapers so I can come up with possible stories. I try to keep an open mind. I try to talk to people. Sometimes, you know, they don't listen. So I've learned to keep my mouth shut too. I'll lose heart for a couple of weeks and just do my job. But then I'll come up with another idea and I'll present it to someone. Not to the top person, but to some of the people lower down. I let them know I have ideas. I think that's a good way to give people the feeling that you're not being used to your potential. It's very easy to be stomped on when you're a production coordinator, you know.

The biggest key is to put in the time. And not lose your self-respect, even though you know you could do more than they're giving you. When I interned at CNN, I was basically a news researcher. I'd pick a story from a newspaper, and if it looked interesting, I'd go to the assignment desk and get the o.k. to develop it. I'd put together more information about it, through phone calls or further research. I was a researcher there one day a week, for a period of four months. It wasn't a paying position, but if you're good, nine out of ten times they'll offer you a job. But you've got to work your butt off. Well, I worked my butt off, but at the end of my internship there wasn't anything immediately available, so I had heard about Financial News Network and I applied. It was in August of '82 that I came on as a vacation replacement. I was basically on a week-to-week basis. I wasn't hired on a full-time basis until a couple of months later. But I had the job, and that was the important thing. The day I started full-time here, I got a call from Cable News Network, saying there was a job available for me there.

It's a wonderful experience working here, even though you can get bored with the actual job. I really want to make it here. At FNN, 50 percent of the people who work here never worked in TV before, so it's a good place for a first-time job. And I think that goes for cable on the whole. You can learn professionalism in cable TV. Positions are created constantly, and positions keep changing. There's room for growing sideways and upwards. We don't have a system down pat yet. So I have the chance to get involved in creating a system. I'm really interested in how it goes on the air. So I watch and I begin to understand. I'm learning as much as I can.

Of course, I want to be challenged more. Sometimes I do feel a little frustrated. But I know that I'm in, and I've got the chance to grow. And I think I'm respected for what I do. One of these days —if I keep my mind and attitude on straight—I'll get a chance to create.

You know what? I hope that chance comes sooner than later.

Alan Zapiken
Scheduler

It has been Alan Zapiken's occupation, since the early fall of 1981, to construct the programming schedule at Showtime, the second-largest pay service in America. It is the scheduler's responsibility to find time slots for the cable network's programs, and to ensure every month that the schedule is balanced with the right shows at the right time. The scheduler who doesn't keep his job is the one who positions *Lady Chatterley's Lover* at nine-thirty on Christmas morn.

Zapiken attended New Paltz College in upstate New York, where he majored in English and minored in journalism. He went on to the Newhouse School at Syracuse University for his graduate work. At the time—the mid-1970s—cable television held as much allure for college students as a place that didn't serve beer.

There were about seventy-five people in my graduating class and only two or three took any cable courses. It wasn't even called cable at the school; it was CATV. And it sounded mundane and boring. People weren't thinking of cable in programming terms, and they certainly weren't thinking of it in terms of jobs. When it came to pay-TV, I envisioned some kind of box on top of the set where you had to keep dropping quarters to keep watching.

After school, I figured I had two routes. I could go to a small town and work my way up the ladder to get to a big city. Or I could go to the big city and try to get on the fringes. So I took that route. I decided to come back to New York, and because of my English background, I thought of public relations. I actually wanted to get a desk assistant job at CBS, but it didn't pan out. In fact, it took me almost a year after graduating to find any work.

Finally, I got a job with a small PR company, which specialized in representing tape houses, TV commercial directors, jingle writers. It was a real small outfit. And after I left the PR company,

I got a job at Hughes TV, a syndication company then owned by Paramount. After it was sold, my job was phased out and I ended up doing freelance PR jobs.

At the end of 1979, I got a call from Jerry Kaufer, who had been at Paramount and then went to Viacom, which owned Showtime. They needed a freelance writer, and he had given my name to them. When I got a call from Showtime, I thought they were trying to sell me cable TV. I forgot that Kaufer had told them to call. So I was appropriately belligerent at first. But I did do freelance work for them in the public relations department, which was really small, maybe four or five people. This was at the time when Showtime had just hit a million subscribers. Within a period of a few weeks, the department more than doubled in size and a position opened. I was brought in to help write news releases. There was a flu epidemic that year, I remember, and I was the only person in the department one day. They kept calling me up to see if I could handle anything beyond news releases. They were amazed I could handle more. My previous PR experience really helped. I sort of evolved into a corporate spokesperson type. I became the senior publicist.

Then one morning, one of the vice presidents cornered me in the men's room and asked me if I ever thought about getting into programming. I said sure. He said, "How about right away?" A job had just opened up for the scheduler. I went through a series of interviews to get the job, and I even drew up a sample prime-time schedule. They asked two other people within the company to do it, as well. So we were being tested for the job. One of the other people did a schedule that was too conservative. The other did one that was too radical. I drew up a sample that really was the middle ground. I made a couple of changes to show I had a few ideas. I didn't really know what I was doing, but I came in with an even keel. I think that's important when you're putting together a schedule.

What is your modus operandi? How do you actually piece together a twenty-four-hour schedule?

You have to take each program as an individual entity. You have to say, "Where is the best place for this program?" and "How do we rotate it to maximize viewing?" You really have to maximize the chance that a subscriber can get to the program, without over-doing it. The perception is that there are too many reruns in pay-TV. We like to look at it as convenience scheduling. I try not to duplicate a day of the week in a month's schedule. In other words, if we're going to run a show six or seven times during the month, I'll schedule it on six or seven different days. We sprinkle it around. You try to find each program its rightful home. If a program runs on a Saturday night at 9 P.M., it won't run again on a Saturday at nine. I do an action plan every month. The twenty-fifth of every month we have a programming meeting, where we present the upcoming shows to every department. We talk about how to promote it, and how promotable each show is. We review each show one by one. After the meeting, we put together a com-puterized title list, which includes the name of the show, at least two of the stars, a synopsis of the program, a few reviews, and other pertinent information about each program. Its running time and the rating, for instance. While that list is being prepared, Jim English and I draft a prime-time schedule. It's easier to do a prime-time schedule as a two-man operation. I'm responsible for twenty-four hours a day, seven days a week every month. I know it's valuable—since prime time is the most visible—to work with another person on it. There's a give-and-take; you throw ideas back and forth. This should go here, that should go there. After we've drafted the prime-time schedule, we send it up to be reviewed by the vice president of programming.

I think a scheduler has to have common sense and a sense of responsibility. If we have a program with a G rating, I can put it all over the schedule. If it's an R, it can only show in the evening. There are some after-hours films we run that I'll only schedule after ten. Some, only after eleven. I screen every movie I schedule. The movie *Cat People*, for example, had a lot of gore and nudity and there was even a short fellatio scene, which is a real taboo. We could have run it at eight at night, but we held it to nine. I try to minimize the outcry that you might receive. I'm very careful not to have anything spill over from late night into the morning, when a kid could be getting up and turning on the set. If a film has heavy nudity and violence, I'll schedule it so it cuts off by five or five-thirty in the morning. The point is, you can run risqué programs and you can do it responsibly.

In what other ways can a scheduler screw up?

You can't make mistakes with the running time, that's for sure. If you have a two-hour movie and you put it in an hour-and-a-half time slot, you're in trouble. When I go to the movie theater, I pay attention to my watch. I'm ecstatic if a movie runs for one hour and fifty-eight minutes. Then, when it finally gets to Showtime, I'll be able to put in a short and the continuity will be easy. If I see a film that runs two hours and three minutes, I'm not in a good mood. Then I have to stretch things. You know, when it comes down to it, I'm not a real fun guy to go to the movies with. I'll go to *48 Hrs.* and I'll watch the audience whoop it up, and I'll think to myself, "This is a perfect Friday night movie for Showtime." There are Friday night movies and there are Tuesday morning movies. You have to know the difference.

Once the prime-time schedule is approved by the higher-ups, what's your next step in putting together the twenty-four-hour package?

Once prime time is set, I look at the large-grid calendar that's on the wall of my office. Then I do a couple of different day parts —day by day by day. It can be very tricky. You have to target different programs at different times of the day. Classic movies run at a certain time. Family films run at a certain time. Then you have to rotate them, so people don't think that one show is on too much. By the seventh or eighth of each month, the entire schedule has to be delivered to the people who make the guides.

When all is said and scheduled, how do you get satisfaction out of a job completed? Where do you derive the feeling of pride in your work?

I take pride in putting it all together. I take pride when it's not changed much by the people I report to. I consider structuring a twenty-four-hour schedule a real accomplishment. And the fact that we're continuing to grow, and there's not a lot of churn, is a reflection that I'm doing my job. A good scheduler has to be detail-oriented. You have to have an idea of why things work together, and how to establish a flow. You also have to like television. I enjoy the fact that I ended up doing what I always wanted to do. I mean, my parents always used to complain that I watched too much television. Now it's "My son, the cable TV scheduler!"

Jane Taylor
Affiliate Sales Representative

Affiliate sales reps need sturdy luggage and a built-in immunity to airline food. Jane Taylor, for instance, travels 75,000 miles a year on behalf of the Nashville Network and Satellite News Channel. As affiliate sales representative for Group W Satellite Communications—which owns a large chunk of both cable networks—Taylor is responsible for the northwestern section of the United States, primarily Oregon, Washington, and Alaska. Her mission is to sway cable systems in that area into accepting the networks. If she's successful, Taylor is then responsible for making sure the Group W cable networks are properly promoted, marketed, and handled by the cable system—or affiliate. An affiliate sales rep is the mouth—and conscience—of the cable network on the road. Jane Taylor has no fear of flying. She's part Willy Loman and part Lewis and Clark.

I had the chance to get the Hawaii territory, but I chose Alaska instead. Alaska is a new frontier, an unknown territory. Like a lot of the Northwest, it's beautiful country and it's hard country at the same time. I was born and raised in Portland. I know the North-west. Some people thought I was crazy, but I think selling cable TV—and our cable networks—in Alaska is a great challenge. There are also a good many country-music fans in Alaska. I guess the reason I got into cable was for that same sort of frontier appeal. I had been a public relations director of a shopping center in Portland, but it was basically a maintenance position. They had been doing the same advertising and promotion for eighteen years. Then I worked at an advertising agency and I was gone after three or four months. I was so bored. I didn't want to work for $700 a month for the next five years and maybe work my up to an account executive position. I wanted to find something where I could get in and prove myself. I wanted to find a job that was moving forward. I didn't want to go through some prescribed ranks and end up nowhere. I didn't want to stagnate. After I quit the ad agency, I was doing some temporary secretarial work just to keep the bills paid. Then one day, I saw an ad in the paper for a Liberty cable company; they were looking for a customer service person. That didn't sound too great, but cable TV sounded intriguing. I knew nothing about it—absolutely nothing—but I called them up and said, "What do you have?" I was referred to the marketing director, who told me they had sales positions. I thought, "Oh, no, door-to-door sales." But when I came into the office to hear more, he got me enthused about the industry. He sort of blue-skied it, but it sounded like I could get hooked. Then he said, "I've never had a woman be successful in selling cable TV door to door." I told him, "Fine. That's a challenge that I can handle."

My job was to go into homes with my blue jeans on and a T-shirt and a little wrench in my back pocket and a convertor box in my hand and say, "Hi—the office sent me. I have to hook this up to your TV set. It'll just take a moment." You see, we were converting an old twelve-channel system to a thirty-five-channel system. For another dollar a month, the customers got all the new basic chan-nels. Then I'd say, "By the way, do you want to buy Showtime?" It was real easy and real lucrative. But I also realized I didn't want to do door-to-door sales forever. I started to read about the industry, and I knew a lot of things were happening.

About this time, I met a fellow who I wanted to spend more time with, and he lived in Tacoma, Washington. So I moved up there and got a job at a cable company in Puget Sound.

How did you hook up with Group W?

I got involved with the Women in Cable organization, and the president of the local chapter was the administrative assistant to a man named Craig Chambers, who was with Group W in Seattle. Craig was moving down to Los Angeles, said he needed someone in the department down there, and she suggested me. As a born and bred Northwesterner, I wanted to go to L.A. like I wanted to cut off my right leg. But after talking with him, it sounded so positive I couldn't pass it up. So I went to the Group W base in Los Angeles, and I ended up getting the Northwest territory. So I spend more time up there anyway. A lot of the time I'm on the road, I stay at my parents' house in Portland, or with my boyfriend in Tacoma, or with my sister in Corvallis. I'm thrilled that I got the territory I did.

Outside of waiting at baggage claim, what does Jane Taylor do for a living?

The job description is I sell and market our service in my territory. What that means is that I sell—or try to sell—our networks to independent cable systems and small MSOs; the major MSOs are handled by the national accounts group. Once someone has bought our networks, I'll sit down with them and say, "What do you want this channel to do for you? Do you want good community awareness? Do you need good PR?" Once the cable operator identifies what they want, then I go through with the planning of how we're going to promote the channel. With the Nashville Network, I've done *everything* to promote the channel, including judging a huge chili cook-off.

You used chili to promote the channel?

It was a big event to kick off the network. I had to judge twenty-six different types of chili; actually, I only had to eat thirteen. Someone else ate the rest. We chose our favorites. We've also done square-dance exhibitions to promote the channel, country cable fairs, nights in the local honky-tonks. Now that I think of it, I'm really glad I like chili and country music.

Chili I can understand. But country music?

I really do love country music. It's a big help. Not all of our affiliate reps do. But it's a piece of cake for me. I walk in and I'm selling a product that I love to the cable system. Here I am—upscale, college-educated, a reasonably good income—all that stuff, and I love it. So I can point to myself and my friends as examples of the people who listen to country music. And I can point to Charley Sixpack who works at the factory.

How do you sell country cable and Satellite News to cable systems? What's your approach?

If I'm doing a presentation to the regional manager, and he's got all the system managers around him, I've got slides, and printed handouts, and a tape that gives an example of the programming. If I'm meeting with Mr. I've-Owned-a-Cable-System-for-Twenty-Years-and-Don't-Try-to-Impress-Me, I still might bring the tape along to show the programming. I do the rest on my own. I'll say things that I know will get a reaction. If a system manager says, "I thought Nashville Network was just going to be a country MTV," my response is that country music is not the same genre as rock 'n' roll. Rock is a fantasy medium. Country music is a real, day-to-day medium. I say, "You wouldn't be happy watching Merle Haggard smash a guitar over the head of a panther." That's a buzz phrase. There are a couple of systems I have which cover large ethnic neighborhoods. One in particular has a large Vietnamese population, and they're having trouble selling cable to them. So I said, "Great. Now you can tell them you have country music. You don't have to tell them which country!"

What's the toughest part of your job?

On the personal side, the most frustrating thing is that with the amount of travel I do, I don't have a home life. I'm real fragmented. I'm in all these places every week and it's tough sometimes to keep up being positive and warm and outgoing when you don't have anything to nurture you on an ongoing basis. On a professional level, the toughest part of the job is closing the deal. Asking for the order. As a little girl growing up in the fifties and sixties, I didn't learn how to do that. I learned how to avoid that at all costs. Although I've learned some good techniques for it, it still gives me a twinge anytime I have to do it. I want to be a nice girl. I want to say, "I understand that you don't have channel space and I really sympathize with your franchise problems and—God help me—I don't want to make you unhappy with me." With *me*. That's about the time when I have to pull myself out and think, "Jane, it's not you. They can like you and not like your product." And I'm growing past that, too. The other day, someone ended up buying one of our channels and not the other and he said, "But I want you to know you did a good job." And I thought, "If I did a good job, you'd have that second channel on." So I'm starting to measure success differently now. It's not in people thinking that I'm such a smooth, warm, comfortable person. It's that I want to get the sales. It's "When do you want to launch this thing?" That's much more important than what they think of me as a person.

Was there a turning point that changed your thinking?

The particular moment came, I think, when I went back to one of the cable systems where I used to work, and went in to pitch the Nashville Network. My former employer was not excited about country music at all. "It's a flash in the pan. *Urban Cowboy* was three years ago and now nobody's buying cowboy hats. What makes you think this is going to stick around?" I took a deep breath and I took out a little pen. I said, "WSM (Group W's partner in the Nashville Network) was formed at the turn of the century. They founded the Grand Ol' Opry in 1920. They own Opryland Communications, which was formed in 1970. The Nashville Network is the only other thing they're involved with." I drew a little organizational chart and I said, "This is not a flash in the pan." And she bought it. And what made me feel like I had matured as a salesperson was that this was a very powerful, very intimidating woman sitting across from me. And she didn't intimidate me enough to make me forget my facts.

What kind of person does well as an affiliate rep?

You have to be positive, and you have to be resilient. You have to have a sense of exploration. Curiosity. There is no straight path to becoming an affiliate rep; we had eight affiliate reps here to start with and four were men, four were women; four had cable experience, four didn't; one came from straight out of school, one had worked for the American Ophthalmology Association, one had worked in radio sales. It's not the experience you've had but the person you are. What people hire in an affiliate rep is a positive image. You have to learn to project it.

As far as I'm concerned, this is the place to be. You can really contribute. I get enormous satisfaction out of coming back to the office and saying: "I did it! I used that close! It worked! Who cares if the chili wasn't that great?"

Michael Fuchs
President, HBO Entertainment

The name Michael J. Fuchs is associated with made-for-cable clout. At thirty-seven years of age, he is the president of the HBO Entertainment Group and the executive vice president of programming. He has also been called the most influential man in all television. For those who traffic in such descriptions, let it be said that his power isn't titular. Michael J. Fuchs is the man who makes the decisions that show up on the smallish screen. HBO's importance is so pervasive that Michael Fuchs's decisions show up on bigger screens too. He is a vigorous negotiator, a programmer with principle, and a round-the-clock elocutionist for the cause. There are those who will say, not for attribution, that he has a touch of impudence too. Those are the ones who are used to hearing the word "maybe." Michael Fuchs makes his living in "yeas" and "nays." He came to Home Box Office in 1976, when it was considered by some to be a plague. Prior to the move, he spent eighteen months with the William Morris Agency. He had received a BA in political science from Union College in Schenectady, New York, and a law degree from NYU. Between graduation and William Morris, he worked in two entertainment law firms and served as an advance man for the political quests of the Edwards Muskie and Kennedy. Now Michael Fuchs is the key advocate, arbiter, and architect of the programming on HBO.

When I first joined HBO in September of 1976, I had a gut feeling that HBO could be very, very successful, that this could be a very important business. But when I came here I didn't say this was going to be what it's turned out to be. I think it's happened a little faster than I suspected. HBO's impact on other businesses —the network business, the motion picture business—is more profound than I would have anticipated over this short a period of time.

Basically, I was just glad to be here. It wasn't that I had a million offers and that I said, "Well, I'm going to risk it at HBO." Whether they made it or not was not critical to me at the time. It was simply that I figured I'd just try something else. I didn't do an analysis and say, "It's going to be pay-TV and I'll work for pay-TV." People come in now and say, "I want to be in cable. I want to be in pay-TV." That was not my thinking. I wanted to be in the entertainment industry and this looked like a new, interesting facet of the business.

Nothing I was doing up until that time in 1976 was very satisfactory to me or very challenging. I've been a lawyer, then I worked at the William Morris Agency. And I had had some contact with HBO. I think I liked the idea of a small, struggling business. Psychologically, that worked quite well for me. And HBO was certainly a small, struggling business in 1976.

What did you do at the William Morris Agency?

I was primarily in what's called the Business Affairs Department, which was then comprised of lawyers and negotiators on their way to doing something else. It was essentially luck that I found myself at HBO.

Luck in what way?

If I had come here a year later, I'm not sure I would have had the same job. I mean, the guy who tipped me off on the job was offered the job before me. He took another job; he said this job had too much programming, which, to me, is like saying there's too much chocolate on the plate. So there were a lot of fortuitous circumstances in getting the job. If I was walking in today, I wouldn't be qualified for the job that I have now. I wouldn't have been qualified then, either.

And today you're called the most powerful man in television.

It's not true. The standards for measuring power are the amount of money you have to spend. And HBO doesn't spend the amount of money that any individual commercial network spends. Power is spending ability. I think there are people who can sit down and think and project that we will. We have more influence in the motion picture industry than, for instance, the networks do. We're spread a little thinner than the networks, in terms of the industries we work in. HBO is a bit of a hybrid. One of the fun things about working at HBO is that one night you're in a nightclub, the next night you're dealing with motion picture executives, and then you're working on a comedy show. If you're a generalist, which is how I view myself, you have ample opportunity to be a generalist. But my power is exaggerated.

How do you describe your power?

Well, I think primarily a lot of people in the industry see that power coming from the fact that HBO is very much involved in a kind of movie financing—its pre-buy activities. The pre-buy activity is the difference between having movies made and not made, and I think, in a sense, that's the ultimate power in Hollywood— the ability to get a project done. And the biggest projects are movies. To get a movie done is considered to be a rather important contribution to the industry, a rather important event. So I think from that point of view, it's HBO's potential influence on getting the movies made and how that affects the future of the motion picture industry. Will it be an industry that produces for the home, or will it be an industry that produces for the theater? Will it be a pay-per-view industry or will it be a subscription industry? Will it be both? I think that the people who make decisions at HBO are, in a sense, important contributors to the process and to what's going on.

What is it about your job that you enjoy the most?

The diversity. It's a job that has me heavily involved in the creative side of this business, as well as the pure business side. That's very good for someone like me, who likes to jump around. I used to spend more time working on the shows at HBO, which I like very much. I don't do that as much now. But every day is something new. It's being part of a new, growing business. It's being part of history. A continuing level of excitement is delivered by that fact. It's kept this job interesting from day number one. I've been exposed to different things that I didn't have on other jobs. It's been a very unique, interesting experience. And, you know, it's an enormously competitive business and I'm a competitive guy. I get a lot of psychic gratification out of that. This place hits a lot of the needs I have for an enjoyable career. You know, I wouldn't have made a very good foreign diplomat. I'm made a little differently than that.

Michael Fuchs will never be an ambassador?

Basically, I don't have a lot of patience to waltz around. This business is a fairly confrontational business. People call you up and say, "You've got twenty-four hours to give me an offer." It isn't: "I think we should talk about this for a few months and sit down, put a task force together, and look at this canal treaty for a couple of years." This is a very fast-paced business and it's a decision-making business. One of the things that's surprisingly missing from a lot of programming executives is the ability to make decisions. I'm not even saying the right or wrong decision, but *any* decision. You know, what's good, what's bad. And that's something that I really like to do. I walked out of *E.T.*, for instance. A friend of mine said to me, "That's going to be the biggest movie of all time." And I said, "You know, I don't think it's got enough going for it to get any type of repeat business. I don't think it's going to have a phenomenon attached to it." Famous last words. I know I'm never going to be right all the time.

What didn't you like about E.T.?

It was just a little trifled for me. You have to understand: I get enormously involved in a movie. I mean, I came out of *An Officer and a Gentleman* the other night and I was, you know, in love like Richard Gere. I really think I'm in this business because I do get enormously involved. Even as a kid, I did. I guess it's a kind of escapism. When I first got here, I said that HBO, this little floundering network, was the greatest Erector set on the market. I never actually had an Erector set when I was a kid. The first thing I do in the morning when I get up is turn on HBO and Cinemax. Just a split second each, just to make sure they're on. You have this living, breathing network that you're involved in, and it's on in your house and millions of other people's houses. And to me, as a child of television and a child of pay-TV—in terms of my business experience—it's still a great thrill to me."

How does it feel now to be the big kid on the block?

I was here when we didn't have that size and we didn't have any muscle. And a bunch of people built this company. To me, business is business. I'm going to do everything I can to keep this company what it is. We happen to be more successful than most people. I think that's something we should get some credit for, and not resentment. But there isn't much loyalty in the cable business. No one's happy till you're out on your ass. I think it makes good copy in a newspaper like the *New York Times* to say, "Listen, we're all smarter than these imbeciles and all they run is fluff. We are much more serious, intelligent people than these executives at HBO, who are lawyers, MBAs, and accountants." The *New York Times* has a TV columnist who doesn't like television.

So I hear things from the entertainment press that one day has HBO losing subscribers all over the place, and the next day, HBO is dominating the world. I hear that we might get into buying Arab countries. The press likes to dramatize. They're all frustrated screenwriters. I believe a good many cable operators are underselling HBO, too. The cable industry feels: "Hey, we don't need HBO. HBO is a nice service, it does well for us, we'll take care of the rest." They're more interested in selling everything else than in selling HBO. Well, they'll learn very quickly. It's the glamour and excitement and the programming strength that a channel like HBO delivers which makes the business. HBO is the thing that can move the cable business. If a cable operator doesn't take advantage of that, he's cutting off his nose to spite his face.

Ted Turner relates a story that Bob Wood, former president of CBS, told him. Wood said that whenever he walked through the doors of CBS, he'd leave his conscience outside. Does Michael Fuchs leave his conscience outside the doors of HBO?

Television is an industry which is quick to exploit, and where exploitation means profits in most cases. I think that maybe the most important contribution we've made is to have a television entity which sets a certain tone and has certain principles and a certain integrity. I think that when you're coming into the home, which is a very intimate place, and you're charging people on a monthly basis, there's got to be some long-term relationship there if you're going to have a business. A long-term relationship has got to be built on the same characteristics on which long-term relationships are built between people: honesty, loyalty, integrity, trust. That might sound a little too subtle—or a little like crap coming from a television programmer—but we really believe in it. This is not a rip-off, exploitation company. And I think HBO has the strongest level of loyalty in the entire television world. Yes, I'm able to do my job, which is not an easy one, and I still keep my conscience with me when I walk through the doors in the morning. That's almost like a dream. I don't think there are many television programmers who can say that.

SIX

The Supporting Services

Career opportunities are not limited to those within the cable companies. Cable television has provided a market of seemingly infinite potential for equipment manufacturers and suppliers, capital venture firms, program makers and program distributors, consulting groups, marketing organizations, and other companies that provide service and software to the cable industry.

The supporting services are the life-support systems that keep the cable industry in the pink. For those seeking a career in cable TV, they represent a "hidden" job market ready to be explored. The supporting services, the facilitators behind the scenes, can greatly expand your career options.

THE SUPPORTING SERVICES JOB FILE

Advertising Madison Avenue is about as close as you can get to Hollywood without acquiring a tan. And advertising agencies in the major Apple—and in centers such as Boston and Chicago and L.A.—have begun to come to terms with cable TV. Some of the more progressive agencies have even begun sinking client dollars into the ad-supported cable networks. Conversely, a number of the more established made-for-cable networks have hooked on with advertising agencies in order to promote their service. The agency will then create and place commercial spots for, say, Home Box Office, to be seen on other cable networks, the broadcast networks, and in print.

Another area of growth in advertising is the production of commercials for the small business aiming to advertise on a local cable system. Currently, there are 450 local systems that run commercials on LO (local origination) channels. According to Ed Dooley of the NCTA, that number will be in the thousands by 1985. With more small businesses turning to cable for exposure, there will be an increasing need for production people to make commercials on a low-budget basis.

Advertising agencies employ a variety of backstage personnel, including the account executive (a liaison between the client and agency), copywriter, art director, producer of TV commercials, media director (who researches the best outlets in which to place advertising), and office workers (secretaries, receptionists, etc.). Several of the giant agencies also employ specialists in the new media, forward-thinking souls who are responsible for keeping a bleary eye on cable, home video, the alphabet soup of modern TV (STV, MDS, DBS, etc.), two-way technology, teeny TVs, projection television, and anything else that people at home will pay to see.

Advertising Sales Groups/Cable Reps   A recent ad in the job opportunities section of the Los Angeles *Times* read: $1000 WKLY in ADV. SALES/CABLE PROGRAMMING. OFFERS A MOTIVATED, MODERN THINKING PERSON A CAREER OPPTY. & UNLIMITED ADVANCEMENT. COMM. & BENEFITS. Sherwin Communications, like numerous other groups across the wired terrain, was commissioned by a cable network to pound the local pavements in search of advertisers and the checks they write. The independent sales companies then hire salespeople with hardy soles and hearts of stone—those are the people who will try to get the sponsors. "It takes an unusual breed to sell cable to the spon-

sor," says Sherwin's Stan Snider. "Some people think you need a
head of stone to get involved. But the people who will go through
the growing pains with cable—they're the ones who could end up
making it big."

Snider said he received three hundred responses to his ad and
ended up hiring only two—"You have to go through a lot of people
before you find someone who can sell cable TV." Salespeople for
independent sales groups like Sherwin usually make a meager salary, approximately $200 a week, and hope for the best in commis-
sions. As the dollars invested in cable advertising grows (a projected $70 million on local spot sales in 1983), so will the
importance of the cable rep firms, which represent cable TV's
advertising opportunities to potential sponsors with dollars to
spend. The oldest cable rep firm, Eastman Cable (established in
1979), has four regional offices across the country (Chicago, Atlanta, New York, and Los Angeles), and currently employs thirty
people. "Selling local and regional cable spots to national sponsors
is still a very difficult sale," said Angela Pumo of the New York
office. "The biggest problem is the lack of numbers (viewer ratings). Still, I think this is going to be a very big business. I'm betting
the best is yet to come."

Brokerage and Financing   A brokerage house will represent cable
systems that are up for sale, and try to match the system with a
buyer. A broker will also assist the buyer of the system with financing. A cable system, after all, is an expensive proposition.

Cable Commissions   To date, eleven states have adopted legislation establishing various types of state bodies that regulate the
cable industry. (State regulatory bodies governing cable exist in
Connecticut, Nevada, New York, Rhode Island, Vermont, Alaska,
Hawaii, Massachusetts, New Jersey, Minnesota, and Delaware.)
The cable commissions provide a liaison between cable companies
and local businesses, government agencies, and labor unions. On
the broader scale, the National Cable Television Association

(NCTA), which employs sixty-two people, the National Cable Television Information Center (NCTI), which employs fifteen, and the Cable Television Administration and Marketing Society (CTAM) represent the cable industry in courts of law, before state regulatory agencies, and with other industry groups. Major emphasis is placed on keeping their members and the public up to date on new cable technology, programming, and the panoply of services that cable television can offer. Jobs at these various associations and commissions range from director to public affairs representative to office staff workers to researchers.

Cable Magazines   Trying to find a decent cable TV listing is enough to turn strong men into cooked manicotti. The problem is that the proliferation of cable networks means more shows to list. And, more importantly, the markets have changed. In the New York area, for instance, more than sixty different cable systems offer service, none offering the same combination of basic networks (ESPN, USA, MTV, etc.), out-of-town independents, pay channels (HBO, Showtime, Spotlight, etc.), and local-access channels.

In response to the cable invasion across the country, *TV Guide* —the granddaddy of the "what's on" magazines—has vastly increased its cable listings. Others, including *On Cable* and *Cable Today,* have positioned themselves for a chunk of the cable market. Each of these magazines not only contains cable listings but also features stories about the personalities who make the channels click.

Cable magazines employ a wide range of personnel, including writers, story editors, photo editors, marketing people, circulation directors, listings editors, and a variety of office staff. In addition to the cable magazines, which detail program schedules, there are a number of cable-related publications, among them *Cablevision, View, The Multi-Channel News, Cable Business, Cable Marketing,* and *Cable Age.* These magazines also employ a plethora of people, including critics, columnists, and entertainment reporters. Since the cable publications stay on top of the industry, a few knowledgeable insiders end up making the transition to cable network jobs. Barbara Ruger, who worked for *Cablevision,* now heads the marketing of the Disney Channel. Alan Levy, formerly of *The Multi-Channel News,* now works in corporate publicity at HBO. For more information, see the interview with Eric Taub, a reporter for *Cablevision* magazine, on page 186.

Cable Shop Personnel   Also called "convertor stores," these retail outlets are utilized to sell cable's wares and have received their principal inspiration from AT&T's Phonecenters. Since it can seem like a lifetime—yours—has passed before many cable systems pick up the phone—and you can see your toenails grow while waiting on hold—many cable operators and would-be subscribers agree that it is often more convenient to drop by a store to place an order. That's where the cable shop comes in. Not only will a potential subscriber avoid waiting on the phone, but he/she will get to watch samples of cable programming while being asked to subscribe. Another primary benefit of launching a cable store is increased public awareness of cable. Cable shops, which showcase cable service, usually employ from three to five people in each location. These are the people who can answer questions and assuage your guilt. What guilt? How about the story of David Cobb? "My wife said ESPN was the reason she divorced me," said the thirty-one-year-old Texas native. "I guess she got tired of competing with cable television sports." For more information on the cable shop employee, see the profile of Dan Bowen on page 184.

Construction   When it comes to breaking new ground in cable television, the construction companies get the first dig. Many independent companies specialize in the design and construction of cable systems from the start, which is called the "turnkey" process. Once a cable company gets the franchise and receives the permits to build, it will work with an independent construction company on the blueprints and, in turn, the construction group will hire the necessary personnel to build the system.

Construction companies with cable experience also provide service when an operating system wants to rebuild and extend its territory. Construction titles include the laborer, crew chief, entry-level lineman, senior splicer, installer, and construction supervisor.

Consultant   One sign of a growing industry is the number of consultants making a living in it. Cable television abounds with consultants of every description, including those who legitimately have something to offer. Consultants are the people who are hired on short-term schedules by cable systems and cable networks to supervise operations, plan marketing moves, assist in the training of customer service personnel and salespeople, provide feasibility studies determining a franchise potential and value, create techniques to increase system revenue and/or system size and scope and, as a service to investors and lenders, produce in-field operational and engineering analysis performed in advance of a purchase or loan. There are general consultants, legal consultants, market research/advertising consultants, product sales consultants, program consultants, and people who just like to call themselves consultants.

Distributors   There are two types of distribution operations in the cable business, studio-owned and independently owned. The studio distributors lease only those films or programs produced by their own studio. That is why a Twentieth-Century Fox picture like *Star Wars* is distributed to pay-television by a Fox department set up to sell to that market. The independent distributors see all available films, attend film festivals, and purchase films to distribute. Both operations distribute to various markets—known also as windows—such as foreign and domestic theaters, network and syndicated TV, overseas TV, home video, in-flight entertainment, and pay-TV/cable. In many cases, distributors also develop product for the cable television market in association with producers who bring in the ideas. For more information, see the profile of Mort Marcus on page 178.

Door-to-Door Sales Groups A cable system might hire an outside sales group to market cable service door to door. These sales organizations hire full-time and part-time personnel to do the knocking. They are responsible for teaching the employee the cable pitch and the ways to reach the "untouchables"—the people who have already been offered cable service and turned it down.

Equipment Manufacturers and Suppliers There are dozens of companies that make the hardware and then, as suppliers (vendors), sell the products to cable systems. What kind of cable equipment does an equipment manufacturer make? How about: earth stations, set-top terminals, amplifiers, splitters and splicers, cable-cutters, tap-offs, wall plates, drive rings, antennas, transformers, plastic wire ducts, underground marking tape, winches and hoists, modulators, signal processors, crimping tools, bonding clamps, drill bits, pad locks, attenuators, tap rackets, ground straps, ladder racks, grounding blocks, drop connectors, digital multitesters, directional couplings, convertors, coaxial cable, a/b switchers, diplexers, and even nuts and bolts?

There are equipment manufacturers and suppliers who make studio equipment, too, including cameras, lights, and character generators. A sales job in such a firm can be good experience for someone pursuing a sales position in cable TV, or it can be a good career in itself. Equipment designers, of course, need a thorough technical knowledge and an understanding of cable television's technological trends.

Production Company Many independent producers maintain small companies, which employ receptionists, secretaries, and staff workers, including program developers, co-producers, and business affairs personnel. The independent production company will more than likely develop and produce shows not only for cable television and pay-TV but also for the commercial networks and possibly motion pictures as well. For more information, see the profile of Chuck Braverman in Chapter 4, The Made-for-Cable Networks.

Recruiters A number of executive search firms and professional placement groups have been formed to supply personnel to the cable television industry. Some maintain computerized job banks in order to place candidates in cable system jobs—and cable network jobs—across the country. Jobs at these recruitment and placement centers range from career consultants to senior supervisors to office workers, all of whom must have a working knowledge of the cable television business. After all, how can you place a client in a job if you don't know what the jobs are? Many of the placement centers do a very good business in matching chief engineers, technicians, installers, and marketing personnel to an industry that needs qualified people.

Teacher As the cable industry has bloomed, so has the need for educators who know the field. Many colleges and universities offer programs that relate directly to the cable boom, as do a number of vocational centers across the land (see Chapter 7, An Education in Cable TV). Hence the need for instructors who have proven cable backgrounds and the ability to translate cable-chat to job-hungry students.

SUPPORTING SERVICES PROFILES

Mort Marcus
Distributor

Pay-television provided a chance for Mort Marcus where there wasn't a chance before, and he has made the most of it. At twenty-nine, he is considered one of the brightest young salesmen in the business. As head of pay-TV sales for the Samuel Goldwyn Company, a television and film distribution firm based in Los Angeles, the peripatetic pitchman spends his days on the phone, running to meetings, and chatting it up at conventions usually opposite buyers from pay-TV networks, who determine whether they will say yes to what Marcus is selling or pass.

Mort Marcus is also responsible for sifting through new program ideas, which are brought to him by producers, writers, and assorted creative types in search of money, the credibility of the Goldwyn name, and a sale in pay-TV. Since the pay-television market has expanded—and with it, his responsibilities at the company—Mort Marcus hardly has time any more for Frisbee football, a passion that rates in the same group with the San Francisco Giants, Manhattan Beach, and an okay from HBO.

My job is very, very—what's the best way to put it?—*horizontal.* That doesn't mean that I lay out at the beach and take meetings in the sand. It's horizontal in that I have a whole bunch of areas that I work in. The vice president of pay at Paramount, for example, is responsible for selling the movies that his theatrical department gives him. He sells to HBO, Showtime, the Movie Channel, and that's basically it. He may seem busy but he only makes two deals a year; he comes in at the first half of the year, then he waits a few months and comes back for the second. Usually, he will wait until one of the pictures is a hit, so then he has something to package the rest of the movies around, and it increases the asking price.

My job is to sell our movies to HBO, Showtime, the Movie Channel, and all the others, too. I do all of them. I cover everything in the pay market. I'll deal with HBO, which has 12 million subscribers, and I'll deal with a system that has twelve subscribers. Paramount, and the other big studios, also have a vice president of development. That person sits around all day and takes meetings to try to figure out a few shows and specials to sell to HBO and Showtime. I do that, too. I develop the programs we try to sell.

Further, I'm also involved in deciding which movies and shows to buy from outside sources. Our acquisitions department gets a hold of a film and they say to me, "Mort, we have a screening of it at three o'clock tomorrow." I watch the picture, and tell them how much I think it is worth in pay and syndication. We formulate —based on my numbers—what we're going to offer to buy that film for distribution, if we decide to at all. Back to Paramount: their pay-TV division distributes movies that the theatrical division makes and releases. The pictures that I distribute are pictures that my boss and I have determined the value of—and then we have to get that price in the marketplace. We are, in a sense, creating our own inventory of films and shows, and deciding how much it is worth.

At the big studios, they are much more fragmented. On the other hand, the dollars that I bring in are not a tiny percentage. I bring in a heavy percentage of the company's dollars. At the major studios, there are so many jobs that everybody is very expendable. Here, I might bring in 40 percent of the company's sales.

What was your first step toward a career in pay-TV distribution?

I got my degree from San Diego State, and I actually wanted to be a director. In college I had taken directing classes, and it was the only thing that I got an A in. Advertising, I got a C; business management, a C. And I had to fight for those. I graduated with a 2.3 average. But while I was in college, I directed a show called "San Diego Sings," which showed on a local station. I certainly wouldn't want to sell it now, but it gave me the idea of being in television. I really wanted to be in films, because it has a certain class that TV doesn't. But I also knew there were a lot more jobs in TV. Or so I thought.

The first thing I did after graduating was selling time on a local radio station. That tells you how many jobs were available in TV. I started out making $600 a month, and it took me four months to get the job. I had to beg to get the job. It was a small Riverside [California] radio station and it was a very tough sell. I was there for two months, and I was there longer than any other salesman.

While I was working at the radio station and on the street trying to sell spots, I kept in contact with a couple of friends who worked at Vidtronics, which is a tape and editing facility in Los Angeles. I wanted to get a job there. So one of my friends told me to call the head guy in the shipping department and he told me that there was nothing available. But I guess he liked my attitude and he told me to call him every two weeks, to see if something opened up. "You won't be a pain in the ass," he told me, "call every two weeks." So I called every two weeks. Sometimes, I'd wait a week and call him after three. This went on for a number of months. Every couple of weeks or so, I'd be told that there wasn't a job.

I was about at the end of the rope with the radio job when I met a guy on the street who told me he was the manager of a direct-mail newspaper and he said that I looked like I knew how to sell. He asked me how much I was making and offered me $800 a month, a pretty good raise. So I took the job, and was pretty unhappy there. There wasn't enough to do, and there wasn't enough money from the short term or long. So, all the while, I kept calling the guy at Vidtronics.

Finally, eight months after I started calling, he said, "Mort, you called at the right time. You finally got a job here." The next week, I went to work in the shipping department.

What did you do in the shipping department?

I made boxes. They had tons of flat cardboard that was pre-lined so you didn't even have to think about what you were doing. I followed the lines, folded the cardboard, and made boxes. You had to wear gloves on your hands or you'd get cut by the edges. I sat with gloves on my hand at a table, folding boxes and putting labels on them. The boxes would then go up to the dub room, where they made copies of commercial tapes to be sent to different stations around the country. I made about a thousand boxes a month so they could cram the tapes in. I also made a big increase in pay from the newspaper job. Since the job was union, I got $215 a week to start, and after a month, I was up to $237. I thought the money was great. I went out and bought a car.

I then found out that Vidtronics owned a company called Gold Key, which was a distribution company. I heard that they were expanding pretty fast. Since I didn't think I had a great future folding boxes, I kept my ears open all the time. You had to stay alert, or you might miss a chance. Anyway, one day I got a call from my boss, who said, "Mort, can you go down to the Bonaventure Hotel [in downtown L.A.] and stay there for the next two days and help Gold Key set up their suite?" There was a big television convention in town, and companies set up hospitality suites where they can meet people, and watch buyers eat their food. Anything to get out, I thought, so I said sure. What Gold Key was doing at the convention was promoting one of their big shows, and they wanted to fill up an entire two-bedroom suite with balloons. Guess what? My job was to blow up the balloons.

But as I was inhaling and exhaling, and watching the rooms get crowded with balloons, I realized where I was and I decided to make the most of it. So I got in good with everybody there. I even stayed late one night, blowing up the balloons. It was then that I met Jerry Kurtz, the head honcho who owned Vidtronics and Gold Key. I introduced myself and told him that I was helping out. He watched me blow up a balloon. I told him that one day, I hoped to get out of shipping.

So he gave you a job and you were on your way, right?

Wrong, Two days later, I was back making boxes. But I thought it would be a good idea to see Jerry Kurtz again. So I went up to Gold Key, went up to his secretary, and said, "I want an appointment." She was a little stunned. I mean, how many people in shipping go up and make an appointment with the president? They don't do it. And the amazing thing is that he saw me. I don't know why. I think maybe it's because he wasn't sure who I was. I think he forgot that I worked in shipping.

When I came in for our meeting, he said, "What do *you* want?"

I said, "Well, I've decided that I want to work at Gold Key." I proposed that, since I was working the four o'clock to midnight shift in shipping, I could come in to Gold Key at three, Monday through Friday, and work for free an hour a day. I told him I wanted to learn about the company. I said, "I'll file. Surely," I said, "you must need something done? You *must* have people who are overworked." I told him that I would even put *boxes* away. And I said that I was interested in sales and that maybe, one day, something would come up. Then again, maybe it wouldn't.

He told me to go see the office manager and see what she had for me to do. Sure enough, I started filing and putting away boxes. Mostly what I did, though, was talk to everybody. I was very conscious of making a good impression. Anything they had me do, I did it well. And everybody liked me. About three weeks later, Jerry Kurtz called me into his office, and he said, "I hear you're doing a pretty good job over here." I said, "Yeah, I'm working very hard." He said, "I'll tell you what: I'm going to Cannes for two weeks, and when I come back, I think there might be something for you." Three weeks went by and I was starting to worry. Finally he called, asked me to come and see him the next day at two.

When I walked in, he told me he wanted to make me his assistant. "You're not going to walk in with me to meetings," he said. "By assistant, I mean you're going to do anything that I need done. Any bullshit, any running, you're going to do it." This was in May. He told me I couldn't start until July because of a budget consideration. I thought that meant that I would get a pretty good raise. Then he said, "I don't know what you're making in shipping, but you'll get the same amount. You're not getting a raise." I told him it was fine.

On July 1, I started as his assistant. Five months later, he started a pay-TV department and he asked me if I wanted to get involved. I said yes, and then I went crazy. I was now head of my own department and I could meet everyone on my own. I was out and visible. And I was a decent salesman.

I ended up at the Goldwyn Company about a year later. And I think I'm more than a decent salesman now.

What do you tell people who want to get a job like yours?

If I was in school, I'd major in TV/radio and minor in marketing, or vice versa. If I ever went back for a masters, I'd get it in marketing. You see, cable TV is still carving its niche. All of the industry needs more salespeople. It doesn't go anywhere unless you have someone to sell it. That's where the jobs are. I would also recommend that people make their own opportunities and chances. You have to be extra-personable, because you never know who will be in a position to hire. And you should get a job folding boxes and blowing up balloons. In between folding and breathing hard, you'll make the contacts that can make your career. *If you're smart enough to plan it out that way.*

Dan Bowen
Cable Shop Employee

What do you get when you cross cable television with consumer electronics in a high-traffic retail setting? In Greeley, Colorado, fifty miles north of Denver, you get Connecting Point, a store which could do for cable TV in the locale what the PhoneCenters did for AT&T. Cable television faces a marketing challenge in the 1980s and the Connecting Point retail stores could help create new awareness and high visibility for an industry that sometimes struggles to get the word out. The new line of Connecting Point stores was unveiled toward the end of 1982, and initial plans call for 675 stores by 1986. The first store, in Greeley, is situated in the middle of a busy shopping center and is attempting to lure passersby who aren't sure what a Connecting Point is. Once they're inside, they find out fast: the name of this retail game is to sell cable TV. You get to look, listen, and sign on the dotted line. The Greeley store employs three workers. One of them, Dan Bowen, grew up in nearby Littleton, received his degree in communications from a school in Boulder, and now works ten to six, Monday through Saturday, transmitting cable fever.

We have a big display area in the store with eleven TV sets, all of them showing different cable services. Then we have a TV set in the middle that has a demo tape; people can walk in and say, "What is ESPN? What is HBO?" and they can see a blurb about each network. It gives them a taste of what they're going to see if they subscribe. All they have to do is press a button and stare. I think it's a lot more effective than a door-to-door salesman who tries to describe what it is. Half the time, you can't figure out what the salesman is saying. Here, you can walk in and see for yourself.

I think a lot of people trying to sell cable TV realized that it was difficult to get people when they were home. A lot of time is wasted ringing bells or making phone calls to people who don't answer. Now the consumers can walk in whenever they want, see what it is, and place the order. And as far as the people who come in are concerned, we are the cable company.

I know it's going to take a while for people to know that this is the place to come. Right now, we're signing up about an average of five subscribers a day. Which isn't bad, when you consider that we haven't done a whole lot of advertising. I think we have to do more. But the reaction has been good from the people who wander in. We sell video cassette recorders, and personal computers too, but people seem really attracted to cable TV. A lot of them ask to see the blurb for the Playboy Channel.

How did you end up working here?

I read an article in the *Rocky Mountain News* about the store starting up. I called up and said I was interested in coming to work. I was really tuned in to retail work; it sounded like something I could learn and do well in. In college, I had been really interested in cable TV. I used to watch it all the time. So it sounded like a good opportunity, and it is. Cable TV needs a customer service center, and this is it. There's a demand for people who can handle an inventory, and do well with other people. I feel like I'm a representative of cable TV. So I smile a lot. My basic feeling is that the industry, like these stores, is going to go places.

The proof is in the people. They come in off the streets and want to know about cable TV. And they really seem to like the immediacy of the store. It's like picking up a fancy phone at the PhoneCenter and seeing how it feels in your hand. Here, people can reach out and touch cable TV. I think that's a great plus. It's nice to know what you're buying. And people take their television very seriously, you know.

Eric Taub
Cable Television Reporter

It is the reporter's job to monitor the goings-on in the cable television industry, and to unravel potential stories. It is the critic's job to view shows, describe them, and share his critical judgments. It is the analyst's job to pinpoint trends, dissect news happenings, and examine how one piece of the puzzle might connect with the next.

As West Coast bureau chief for the weekly trade publication *Cablevision,* Eric Taub is a reporter, critic, and analyst. He digs and deciphers. He stays on top of an industry and a "state of the art" which can change every day.

You really need an overview. That's the most important thing if you want to write about pay-television and cable. You have to ask yourself, "What does it mean?" Neophytes don't write that well about it because they don't know what to write about; they don't know what's important. You have to think of ideas, but you have to find the angle in the idea too.

I fantasize about being on a panel one day, and being asked how to be a reporter in cable television. Fact is, cable TV has more panels going around than any industry I know. My answer to the question would be that there is no one way to do it. There is no one key. No formula.

I had no training at all for it, really. I sort of fell into it. I was a screenwriter and I was making my living as a story analyst for the studios. You know, reading scripts, writing short descriptions of the story. My screenplays were getting interest, but once they got to the studio or the network, they wouldn't sell. A friend of mine had written a piece about a cable consultant named Paul Kagan, who put out a bunch of newsletters for the industry. After the piece came out in the *New York Times,* Kagan offered him a job. He turned him down, but my friend called me up to see if I'd be interested. He asked me if I'd like to write about pay-TV.

My immediate response was: What's pay-TV? All I knew about it—and this was only a couple of years ago—was that some people I knew had a pay service in their homes and it sounded like a gyp. Who cares if you can get more movies on TV? I had no idea about the differences between cable, STV, and whatever. I was very confused.

As it turned out, I decided to meet with Kagan and he asked me to write a sample piece for him. He actually gave me the questions to ask. I had no idea what I was asking, but I wrote down the answers. He liked how I put it together, so he offered me the job. It was really sink or swim. I ended up editing two of his newsletters. So I guess you can say I swam.

A good reporter has to be concise. You have to be a tight writer. I learned that while I was doing the newsletters. You also have to read a lot about the industry, even if you're a writer yourself. Surprisingly, there are still a lot of solid opportunities for people who know how to write and who know about cable. There are jobs available. The fact is, there aren't too many people around who know about both.

Do you enjoy writing—or just having written?

What I like about reporting is that it's a giant crossword puzzle every day. It's a challenge every Monday morning, when I come into the office, and say, "What am I going to write about?" It's like having a heart attack every week. As to the actual process, I like writing stories where I have a point of view, which is not appropriate in news stories. So when it comes to news writing, I like having written. It's more a craft than an art. But when it comes to the longer pieces—the analysis—I get a tremendous high when the flow is going right, when the rhythm of the typewriter is moving and it's feeling like I'm writing a song. One word can change an entire paragraph, you know. It has to click.

Another high is meeting that Tuesday night deadline, and the one on Thursday afternoon. Using my contacts sheet works for me. I have a list of people I call every week to see if anything's happening. It's good to have that trust.

And the lows?

I'd rather not have to pull teeth to get the story. It's uncomfortable and awkward at times when you have to pound away to get a story. Often I don't blame a person for not wanting to talk with me. I probably wouldn't talk to me, either, if I were in their shoes. Sometimes I feel like a weasel, trying to get something that no one wants to give me. It's also frustrating when someone gives you a story, tells you it's an exclusive, and then you see the same story in all the other trades. And it's a bit depressing when people call you up with an idea for a story, which they say is the hottest thing to hit the cable industry. And then, when you ask them for the story, they say, "Get this: Last night I had dinner at a Chinese restaurant!" When I tell them that it's no story, they scream, "What do you mean? It's the first time I've ever eaten Chinese!" Sometimes I get calls from people who say, "Listen, if you print this story, we'll take out a half-page ad." Or I get a call from someone who says, "Print this story; we run half-page ads in the magazine." You've got to separate yourself from that. You have to say, "You know what? I really don't care."

Do you think you have clout in cable television? Can reporters really make a difference?

Actually, I think I have considerable clout. I'll give you an example. About two years ago, Group W in Los Angeles introduced the Galaxy package to subscribers—a combination package of HBO, Showtime, the Movie Channel, and Z, which is the local pay channel. They arranged a deal with all the pay networks wherein they would offer the networks free for a period of one week, for subscribers who wanted to sample the programming. Any subscribers that didn't want to sample the networks were to be trapped—essentially blocked from receiving the channels by mechanical means. Well, there was no way in hell for Group W to block everybody within a period of one week, but they didn't tell the pay services that. I did an interview with the systems manager who said that, obviously, they couldn't do it and that it would take maybe three *months.* I said thank you and I printed the story with the quotes. Well, the movie studios went crazy. They were licensing (selling) their films to the cable networks based on the number of subscribers, and all of a sudden, thousands of people were getting them for nothing. Because of what I had written, a cease and desist order was issued against Group W, the story appeared in the L.A. *Times,* and I changed the policy of a company.

Another time, I wrote something about the union situation in cable, and I got a call from Minneapolis from a man who said, "I didn't know Group W refused to negotiate with IATSE (a local union) in Los Angeles. They're trying to get a franchise here and I'm on the recommendation board and I'm going to bring this up."

So you see, reporters can make a difference. And for that reason, you have to be responsible in your job. You've got to be very careful about what you write. You have to be aggressive but you can't have a lot of aggression. I know a lot of writers who are obnoxious, and because of that, no one talks to them and no one gives them work. If you understand what you're doing, it's a very exciting job. It's a lot more exciting to write about cable TV than about the furniture business, that's for sure. I think it's more interesting than broadcast television too. Cable is changing so rapidly. It's a real challenge to figure out what's going on.

That's why I think all aspiring reporters should stick to it. Freelance, if you can. Go to seminars. Learn about the industry. Keep aspiring. There's a good chance you can make a difference, too.

SEVEN

An Education in Cable TV

The wheels of Academia grind very slowly when it comes to change, yet cable television is quickly taking root in broadcasting departments of the nation's universities. The explanation is that many students are demanding information about the cable industry, and the information has been hard to come by. "Cable's influence and significance has become so obvious," says Archie Greer, radio and TV professor at Ohio State University, "that you just can't ignore it."

In addition to university programs addressing the cable industry, a handful of training centers have begun integrating cable courses into their curriculum. A few of the MSOs—the companies that own more than one cable system—are also assuming a professorial stance when it comes to the cable business. "Cable is no longer a case of technically transmitting a product to people," says Gary Klein, Group W's vice president of human resources. "It's become a major business venture. And as the industry changes and matures, there is a tremendous need for senior executives on down to entry-level people. And it's partly the responsibility of the cable company to educate. We can't assume that someone else will do it for us."

"Professor, what is cable TV?"

Let's find out where that question is being answered.

Perhaps the most well known two-year program in cable TV is offered by the University of Cincinnati. The course of study involves two years of intensive work in cable management, culminating in an associate's degree in applied business/cable television management.

"A graduate of our two-year program can assume any number of positions in the cable industry," says William Frase, the assistant dean of the university. "The program currently has about seventy people enrolled, and they come from Louisiana and Maine and Arizona. There is a real national flavor to the program. When we first started it three years ago, we set an enrollment goal of twenty people. We had twenty people sign up in a week and a half. And over the past year, our perception has been that there is an increasing need for this program. Ours is a college—and this is a course—that will make people marketable in the job arena over the next ten years."

In conjunction with the program, which covers such elements as cable management, accounting techniques, and the principles of cable transmission, the University of Cincinnati organizes other activities covering the cable field. One of their most successful is a seminar series organized for people already working in cable TV. "There are a whole number of folks in it who don't know a helluva lot about it," says Frase. "The first seminar we ran was on cable marketing and the special needs of the cable marketeer. We had many people from the industry because they wanted to find out more about the job they had."

For people not in the cable business at present but looking for an edge in, Frase recommends the two-year associate degree at the University of Cincinnati. He points out that even if one is not at college age, the program could help direct a person to a fulfilling career in cable television. "The average age in the program is early to mid-thirties," he says. "There are a lot of people out there who are looking at a career change and don't know where to turn. All the insight—and opportunity for learning about cable—shouldn't be geared to the eighteen-to twenty-one-year-old alone. Cable television is an opportunity for a career change.

"We graduated our first class last spring, for instance, and of the nine people who graduated, all are now employed in the cable industry. The average salary they're making is $23,000 a year, which is considerably higher than that of a person who just has an associate degree in accounting.

"We offer a well-rounded, academically based program in cable TV. And I believe the opportunities in cable are fantastic. Warner Amex has the local franchise here in Cincinnati, and they are expecting to *triple* the number of people they employ.

"And through cable, there will be different kinds of opportunities. As cable gets into information exchange, to turning off your lights at night, to linking college to college, to buying your groceries at the store, the manpower need—and the need for people with an education in cable TV—will be tremendous. And people are responding. Every day I get a phone call from someone who has a masters or Ph.D. and who says: 'I need a job. How do I get into cable TV?' "

Another university giving cable credit is Indiana U., which introduced the industry to its curriculum in 1973. Indiana University telecommunications students—853 undergraduates and 44 graduates—study production, advertising, audience analysis, communication theory and management as it applies to the general communications field. Two courses deal specifically with cable. One, offered to students at the senior level, is a beginner's look at the major elements of the industry, including technology, industry structure, program development, marketing, and franchising. The other course, offered to graduate students, focuses on the cablenomics of system operation, legislation, and financial arrangements involved in pay-cable programming. Graduate students also can participate in specially designed seminars, which feature industry representatives who speak on industry-related topics. According to Don Agostino, an associate professor, executives of Indianapolis Cablevision, the local cable system, often participate in the seminars.

"We also provide intern programs for the students, which have proved very popular," says Agostino, who initiated the cable program. "Of sixty-five students who participated last summer, one third were placed in cable-related industry positions. We expect to place a comparable number this year."

Interns, he said, get experience in a variety of jobs ranging from marketing to production and sales. The University of North Carolina is one of a number of learning institutions now incorporating cable TV studies into the broadcast curricula. "It is difficult for students to envision where the jobs are because they know so little about the industry," says Loy Singleton of the Radio-TV Department at the University of North Carolina at Chapel Hill. "But there is a real interest in learning about it." Currently, the university offers a single course on cable—a basic introduction course. "We're really just getting started in cable," he comments. "The faculty is very interested in adding more cable material into their classes."

Syracuse University first added cable to its line-up of courses eight years ago, thanks to Professor Vernon Sparkes, who recognized the potential of the industry before many did. Specific cable courses and general communications courses that include cable are currently offered. In addition, the university's telecommunications school offers interested students an internship with a local cable system owned by the Rogers Cablesystems Company.

An interesting statistic: according to Professor Sparkes, approximately one third of the department's 1,000 undergraduates and 70 graduates seek careers in cable TV.

The University of Texas is answering graduate students' requests for cable courses. It offers an advanced course on franchising and an internship with a local system. Studies in cable and other technologies also are being integrated into the school's standard communications course. "It's only a matter of time before cable becomes a regular and accepted part of most communications programs," said one University of Texas professor.

A number of other colleges and universities across the country, such as the University of Wisconsin and New York University, are offering studies in cable television. It is quite apparent from the enrollment numbers that many students aspire to a cable career, while others' interest in the cable business is ignited through cable studies in broadcast-oriented programs. John Abel, professor and chairman of Michigan State's Telecommunications Department, applies what he calls a broad view to the school's telecommunications program. Its center consists of general courses in broadcast and the new technologies, including satellite communications and MDS (multipoint distribution service). The focus of the courses ranges from audience analysis to the study and dissecting of telecommunications policy. The department also offers three cable courses. Cable Communications is one, and it deals with the technology of cable, satellites, and distribution plans. The course was first offered in 1973; its content has kept changing to stay on top of the changes in the industry. Now it delves heavily into two-way cable technology, and also includes educational segments on programming services, operations, and franchising. Recently, a course was added that concentrates on franchising and refranchising. On the graduate level, there is a course devoted to the study of cable system management—an area of great opportunity over the next three to five years.

The school's students share the faculty's growing enthusiasm for educations in cable TV, according to Professor Abel. "We haven't done anything to promote these courses and interest has skyrocketed," he notes. "Classes that had enrollments of 40 to 50 students five years ago now attract 140 to 240 students."

Approximately 40 percent of the school's undergraduates and almost all of the graduate students pursue cable careers. Many would like to follow the trail blazed by Kay Koplovitz, a graduate of Michigan State.

Koplovitz is the president of the USA Cable Network.

While many of the country's finest universities and colleges have begun to turn out middle-level personnel for the cable industry, there are several vocational-ed centers keeping an eye on the entry-level and technical end. One of them, the Garfield Skills Center in Dayton, Ohio, has been offering cable-related courses since 1975, and its record of placement in the industry as of late has been remarkable.

"We just graduated sixty people from the center and now they're all over the country, working in cable TV," says Robert J. Davis, the director. "You come to the Garfield Skills Center not just to learn; you come here to be placed. And as far as I'm concerned, we can't fill all the demands we get for qualified people. The surface hasn't even been scratched."

Garfield offers three cable courses that provide hands-on training for the adult student. The three courses teach the finer points of being a cable television construction lineman, an installer, or an electronics technician, which is the most advanced class. The technician's course provides the student with thorough information on troubleshooting, changing amplifiers, checking the headend, and other cable-related complexities. Each course requires 910 hours to complete, which translates into 27 weeks of full-time learning. The cost is rather steep—$2,200—but Davis believes it is a small price to pay for a career in cable TV.

"The two cable companies who operate in and around Dayton are dominated by our work force, our graduates," he points out. "I know of cable construction people making $80 per day, plus per diem. We sent fifteen or seventeen people into Baltimore. We've sent graduates into Houston and Chicago. The point is: there has been no problem in placement. If you have the education—and you have the ability to move around—it's a booming business. It's like the telephone. Everybody has to have cable."

The East Bay Skills Center in Oakland, California, offers a similar trio of cable courses to that of Garfield; in fact, East Bay served as Garfield's model. The Center in Oakland has been enlightening trainees about cable over the past twelve years, and director Robert Dabney reports that East Bay turns out ninety cable workers a year. "There is a great turnover in cable TV," says Dabney. "The industry constantly has to be fed people who know what they're doing. I don't see that changing, not when most of the country still has to be wired."

Over the past year, IOCs—Industrial Opportunity Centers—in Pittsburgh and Cincinnati have offered educations in cable TV geared to the construction phase of a system. It makes sense that in areas where new cable systems are building, training centers would offer six-to ten-week courses in construction and installation. Mini-programs have been running in St. Louis and Chicago, among several other major urban markets saying hello to cable TV. Many of these training centers offer cable courses in conjunction with the cable company building the area.

One of the most successful of the skill centers delving into cable is the Dakota County Vocational Technical Institute in Rosemont, Minnesota. According to Jim Staloch, the supervisor, this trade school has had a 100 percent success rate in placing its cable graduates. Word gets around; there is a waiting list of seventy-five for the two-year cable course.

"We graduate about twenty-five people every June, and a good many of our graduates are now chief technicians in the cable field," says Staloch. "The others are working in cable in other technical responsibilities. The encouraging thing is that our advisory committee, which is made up of cable managers and experts, tells us that this need for skilled people will continue to grow over the next few years.

"It's also encouraging to note that, on a percentage basis, only 3 to 5 percent of our graduates had any prior experience in cable before they joined up; 30 to 40 percent had some college experience. The rest are high school graduates who had a good interest in electronics. They were very lucky: some bright school counselors steered them to cable TV."

The cable program runs for 220 days a year. The cost is $3.15 a day for in-state students and $5.50 a day for out-of-towners. Due to the fact that the cable program is the most popular of the fifty different career courses offered, the Dakota County Institute in

Rosemont is adding another cable TV session by the fall of 1983.

Claiming a 92 percent success ratio in cable job placements is the Miricopa Skills Center in Phoenix, Arizona. This Center has offered a cable program for the past three years, and divides its course into two phases: classes for the would-be installer and then for the cable TV technician. The admissions policy is open-entry and open-exit, and the cable course takes twenty-six weeks to complete. "Actually, if you can complete the course *before* that, you may," says Cari Braver, the coordinator of student services. "Our objective is to get people to work." The fee for the cable course is $21.60 a day or $2.70 an hour. Students in cable TV are eligible for financial assistance, which comes from various sources including the VA administration and Voc-Rehab, for inmates in prison looking for a better way of life. "We have an incredible variety of people in this program," says Braver. "And we keep getting calls from the cable companies asking us if we have a new bunch of graduates ready to go to work."

The National Cable Training Center, located in St. Charles, Missouri, 25 miles short of St. Louis, is also readying people to go to work. It offers three courses in cable TV: a lineman's course (the construction of a cable system) runs for seven weeks and costs $900; an installer's course runs fifteen weeks and is priced at $1,-500; and the third is a course for future service technicians, which costs $3,000 and covers a twenty-one-week period. The Center currently counts two hundred students and has had a success rate in nationwide placements of over 80 percent. It has been open since 1980. Other training centers addressing the personnel needs in cable TV are based in Cleveland, Ohio, Washington, D.C., Wadena, Minnesota, and LaCrosse, Wisconsin, to pinpoint a few. (For a complete list, check the Resource section in the Appendix.)

Technicians and engineers are two groups where people are in critically short supply. As the newer urban franchises are granted, qualified and proficient personnel are needed to ensure that the "system of tomorrow" works today. "Technicians are always needed," says Debbie Anderson, personnel administrator at United Cable. "Very few people have experience in cable, so they're trained in the job from the very beginning and then work their way up. The majority of our employees are between twenty and twenty-five years of age. Because of the growth and relative newness of the cable industry, the career potential for employees in this age range is excellent."

Times-Mirror's manager of human resources, Wrise Booker, agrees with Ms. Anderson's contention about the need for technicians and adds: "Some cable companies are seeing the need to establish training programs and centers for technical personnel, because the technicians need to be specifically trained."

That was part of the reason behind American Television and Communication's (ATC) construction of a training center at its Denver headquarters site. The first two programs were directed at installation supervisors and installers. ATC's center, however, was directed at its own employees, while Group W's Cablevision Training unit in St. Louis drew applicants from outside the company as well. Several of the major MSOs have been establishing more lasting ties with the world of education, beyond financing quickie training programs that turn out hordes of construction and installation types.

Warner Amex Cable Communications Corp. has financially supported the University of Cincinnati cable programs, and the company began a scholarship program in 1982 that offers a $1,500 stipend to students seeking careers in cable management. Rogers Cablesystems, through its system in the Syracuse University area, has agreed to sponsor a graduate assistantship in cable studies. Other cable companies are aligning themselves with colleges, universities, and training programs, supplying money, equipment, internships, and scholarships. As cable has grown, so has the realization that the industry has to educate.

Other MSOs are currently offering internships, for people who want an education in cable TV. Cox Cable Company, based in Atlanta, started its intern program in 1980; over 250 students have participated in the past three years. Thirty-two of those interns now have full-time positions in Cox Cable systems.

"The intern programs work for the students and they work for us," says Julian Eaves, the administrator of the company's Equal Employment Opportunity (EEO). "The intern gets a first-hand knowledge of a cable system and the exposure which will help that person to decide: Is this the industry for me? And it acts as a feeder for us. It gives us the opportunity to look at a student for up to six months, and it gives us a presence in the community."

Cox Cable owns more than fifty cable systems in twenty-seven different states, and the internships are usually administered by the local system manager, who determines his/her needs and where an intern might best fit. "The opportunity is really out in the field," says Eaves. "That's where the action is. In Cleveland, a young lady came in through our intern program there and is now director of operations. A young man called up, and we placed him in production at a system in Macon. Now it looks like he's going full time." Eaves points out that among the sixty-one interns currently in the field, twenty-two are minority members and twenty-eight are female.

"Our intern program is for students with high initiative," he stresses; "students who have purpose and an eagerness to learn. We don't put too much emphasis on experience, obviously. If they have good communications skills and the drive, that's the key." Internships in a Cox Cable system usually last for twelve to fifteen weeks; unlike many other programs, the intern gets paid for his work— an average of $4.50 per hour. On the subject of the intern turned full-timer, Eaves says: "We base it solely on the quality of the work. We will convert an intern who has shown that he or she can perform the job assignment." Ninety percent of the interns are college students, while the other 10 percent are high school seniors or noncollege people who have shown a strong interest in and feel for the cable industry. Interested parties can send their résumés to the vice president of human resources or to Julian Eaves at Cox Cable headquarters. "In regard to phone calls," says Eaves, "they can call me." In other words, the veep probably won't take your call, but the EEO man will.

Viacom Cable offers a more limited internship, which can be

more accurately described as a management trainee program. Most of the trainees go to Viacom with an MBA in hand from Harvard or Stanford, and then experience cable television through a "rotational" training program. "The management trainee spends two to six months in each area of the business, accounting, production, marketing, and so on," says Eric Ovlen, the director of human resources. "We have extended the program to people who have a certain discipline and helped them become generalists." The management trainee, which is a paid position, will work in a Viacom system from one to two years. "That person will have a leg up on the others," Ovlen points out. "Cable television is moving into more of a business phase now and the trainee will have a handle on the potentials of the industry."

Multimedia—an MSO that operates systems in Illinois, Oklahoma, Kansas, and North Carolina—offers internships in the production of local programming, which includes LO and public access. Says Cress Gackle of Multimedia, "We gain a mule or a secretary or someone who wants to write scripts, and they gain an edge into the business. We feel real comfortable with each other." Generally, interns work for free to pick up the experience at the company's smaller systems and for minimum wage at a few of the larger ones. Several Multimedia interns have become Multimedia employees.

What manner of man—or woman—does Gackle seek? "I want someone with an outgoing personality. Someone who can get started on his own in the morning and who knows how to work with people. If you have those traits and you're in the latter part of a four-year degree program in mass communications, I'd hire you on the spot!"

Gackle is a firm believer in volunteerism, too. "A person interested in a cable career can come in and volunteer the time," he said. "Interns and volunteers are wonderful, because you can observe them without the threat of hiring and firing. And when a job opening comes up, you can keep them on—and keep the flow and continuity. There is no disruption in hiring a volunteer or intern."

Daniels and Associates, the cable concern owned by trailbrazer Bill Daniels, offers internships in a number of cable systems, including those that serve college towns such as Baton Rouge, Lousiana, Greeley, Colorado, Carlsbad, California, and Ann Arbor, Michigan. "The most important step toward a cable career is the exposure to an actual system," says Jim Ruybal, vice president of human resources. "It's not enough for most people to have a degree in communications and all the buzz phrases ready, like 'cable is the wave of the future.' Far more important is to know something about the day-to-day operation of the cable system, and interns get the shot to learn it."

The most ambitious intern program offered by a made-for-cable network is the one at Turner Broadcasting, which owns Cable News Network and superstation WTBS. The program is open to college students specializing in mass communications or broadcast journalism and, while the internship doesn't pay in cash, school credits can be earned. The intern program, of course, is invaluable when it comes to OJT—on-the-job training.

Ten to fifteen interns work at Turner Broadcasting each school semester, and the number increases to as many as forty during the summer break. The internship lasts for a period of sixty to ninety days. The program is so successful that nearly half of the interns come from outside the state of Georgia.

Melanie Davis, who runs the intern program, suggests forwarding a letter of recommendation from your school, a cover letter detailing how long you would like to work, and a résumé. If accepted in the program, the college student will be responsible for finding a place to live—and for keeping up with the variety of roles he/she will fill. The intern will get a taste of everything a cable operation has to show, from floor managing to running the Tele-PrompTer to filing tapes. There is no guarantee that the intern will find a job with Turner after the internship—and college days—are over. "It's strictly timing," says Ms. Davis. "That's why some people wait until their last quarter of school to intern. They figure that if there's a chance for a job, they want to be available to take it."

Unique to Turner Broadcasting is a position called the video journalist—or VJ, for short. An entry-level position, the VJ slot was created three years ago for college graduates and other previously untested cable hopefuls. "The foundation of this company is the young people that we've brought in," says William Shaw, vice president of personnel. "And the VJ position is the heart of the opportunity. We bring in a student who starts off at a low salary —maybe $9,000 or $10,000 a year—and they get involved in writing, gophering, floor direction . . . and then, in fifteen to eighteen months, they become directors and producers. If that same college kid had gone off with one of the biggies like NBC, it could take fifteen to eighteen *years* to move up to those positions. We offer here rapid promotion and the chance to get exposed to everything in the cable business. After a year as a video journalist, the kid can say, 'Hey, look at me! I know how to wear a lot of different hats!' "

And how does one go about snaring the VJ spot?

"Everybody wants to work for Turner, so it's difficult to get in," says Shaw. "But one of things that a student can do is to be the best in your school. I don't necessarily mean the one with the highest grades. More importantly, you have to learn what's going on. We get a lot of college kids in here who apply and who have an amazing lack of knowledge about cable, and about the world around them. It's important, since we're the company that puts on Cable News Network, to know the news. It's also important to have good writing skills." Before you end up with an interview, Shaw notes, you will have to send a résumé to Atlanta, where Turner maintains his cable HQ. Shaw recommends sending along a cover letter, too.

Unlike Turner Broadcasting, most of the made-for-cable networks do not offer regular intern programs or entry-level learning positions. Several, including Showtime, Satellite News Channel, Cable Health Network (New York office), C-Span, MTV, and ESPN, do integrate the occasional intern into the network structure. Even the Eternal Word Network, based in Birmingham, Alabama, uses an intern here and there. Since it has a full-time staff of only fifteen, there are jobs that need to be done.

The best strategy is to send a cover letter and résumé to the respective personnel directors at each cable network—or at least a dozen of your choosing—and then follow it up a couple of weeks later with a call. Once again, timing is probably the best thing you can have in your favor—or the worst thing going against you. If a particular department at the network needs an extra hand and your résumé lands between the right fingers, you could very well gain an internship at a made-for-cable network.

One of the most important points you can get across in your résumé, cover letter, and call is that you're willing, ready, and able to assume a part-time internship. This will separate you from the hundreds of applicants on file who feel they are ready for a full-time spot. Those are the people with degrees, track records, and shows to show. In other words: experience. If you come up a little short on that end, your education is not yet complete. An internship is your next step toward your diploma from the University of Hard Knocks—department of cable television.

On the system level, internships offer the best shot at an education in cable TV—from the ground floor on up. Group W interns in Los Angeles, for instance, spend fifteen to twenty weeks at the system, learning everything from basic studio functions such as lighting to actually directing and editing local-access programming. According to Robin Gee, the young access coordinator, Group W Cable in Los Angeles will bring on as many as twenty-five student interns per school quarter. In Cincinnati, the Warner Amex Qube system chooses interns from local colleges, and the interns receive academic credit in addition to a valuable "in." Richard Johnston, the employee relations manager for the system, reports that "the interns say they learn more in two months here than in four years of college."

Many interns have successfully used their time and experiences as springboards for breaking into the industry on a salaried basis. Ross Rowe, the access coordinator in East Lansing, mentioned that he got his job because the higher-ups remembered him from the time he interned. Sharon Mooney, who works for the UA-Columbia cable system in San Antonio, says: "I was doing intern work while I was going to school. After I graduated, I came back and talked to them about a full-time spot and I got it—because I had experience at the system." Leanne Larson of Long Island's Cablevision, the system that Dolan built, also landed her job after an internship. Just twenty-three, she graduated from C. W. Post College with a degree in communications and, with school and an internship behind her, returned to work at the system. "The one thing I kept hearing," she says, "was that I got the job because I was familiar with what a cable system looked like from the inside. A little education, you know, goes a long way."

If you're interested in becoming an access coordinator on the cable system level, like Leanne Larson, or want to know just about everything there is to know about cable production, a six-week training program is offered by the National Federation of Local Cable Programmers (NFLCP). The training program involves three phases: a six-day intensive workshop, which covers administrative techniques, the management of an access acenter, and TV equipment selection and use; a four-and-one-half-week internship at successful access centers or systems around the country; and then a three-day wrap-up session. Training programs are offered four times during the year: a winter and a spring session, and two summer sessions. Different host sites are chosen for each session. In 1982, programs were conducted in New York, Massachusetts, Indiana, and East Lansing, Michigan. The cost of the training program is $1,750 per six-week session, which includes instruction, text, videotape, and equipment rental. College accreditation can be gained through the program.

The NFLCP, a nonprofit organization formed in 1976, also runs a number of workshops around the country, which range from training programs for potential access coordinators and producers to the care and feeding of the mobile van. For more information about the organization, see the Appendix.

If an internship is out of the question, and the NFLCP's price is too steep, by all means sit yourself down in the middle of a workshop program offered by the cable system, a community-access group, or even the local library. Most of the workshops are offered free of charge or require only nominal fees for materials. The workshop is a perfect place to learn the basic production skills and a bit about how cable television clicks from the inside out. It is a learning process that might very well pay off.

EIGHT

Breaking In: Making the Right Moves

Job hunting has become an essential human survival skill. The unfortunate fact is that most people have never bothered to learn that skill. There are some people who actually believe that all they have to do is show up at the front door, diploma in hand.

The simple and ineluctable fact is that, in today's job market, finding a job has become akin to marketing a product. The product is you. You must know how to package the product appealingly; how to research the market to match yourself to employer needs; and how, like the card-counters in Vegas, to stack the deck at least a bit more in your favor.

As in other industries, there are no sure bets when it comes to finding employment within the cable industry. It takes the patience of Job, since few people are lucky enough to turn a quick winner. It will take long hours and lots of stamps. But the bottom line is that you should get out of it what you put in.

The cable television industry, *unlike* most other industries, is growing rapidly with varied career opportunities. It is up to you to take that fact to the bank.

Here is a consensus of advice: nine steps toward your first job in cable TV.

1. Fully Assess Your Skills and Aptitudes. Many people already employed in the cable world, and many authorities on employment in general, suggest candid personal stocktaking as the starting point in any job hunt.

From Paul Shay, director of human resources at Showtime: "I think that it is very important and vital that a person assess his or her own talents and feelings about particular opportunities. I think it is very helpful for a person to go *inside,* and find the right direction in connection with careers. Would you rather be standing up in front of a crowd giving a speech or would you rather be writing a speech? It's extremely beneficial to know what you can do—and want to do—before you decide which direction to go in cable, or in any other industry. I think everybody should do a *reality* check. You have to assess your own interest levels."

Shay has been in the cable industry since 1982, and credits the Strong-Campbell interest-inventory test for steering him in the right direction. Occupational skill and aptitude tests, administered by career counselors, can often serve to focus an individual on what it is he or she wants to do. Paul Shay continues:

"The Strong-Campbell test is a fifteen-part, 340-question test that deals with general occupational themes. Would you rather be doing this than that? If you had a chance to crunch numbers, would you rather crunch numbers or go out and play softball? Would you rather be fixing your car or sitting down and eating dinner? It gives you direction. During my job search, I had an opportunity in the banking industry but I knew I wasn't right for it. It's too structured an environment. The test didn't tell me where to go but it told me: is this right or not right for me?"

The key factor is being honest with yourself from the start. Are you really cut out for a career in cable TV? Or are you only attracted to the so-called glamour of it all? Remember: once you are in the cable industry—and this goes for the entertainment field as a whole—the glamour begins to look more and more like work.

From Dwight Tierney, vice president of human resources at Warner Amex, which includes The Movie Channel, Nickelodeon, and MTV: "I want people who are committed to cable television and not just to getting a job. I don't want someone who is just as committed to scraping the grill at McDonald's. I want people who understand the product and the process of cable."

Paul Shay concludes:

"You have to get a handle on yourself. When I went to a career counselor and said, 'Help me,' and after I took the Strong-Campbell interest-inventory test, I knew for sure where I fit. Everybody has basic interest scales and mine were music and dramatics and art and public speaking; in other words, this industry. The test also revealed that I could be in the restaurant business, because you deal with people there. My feeling was that the restaurant business was not on the leading edge of technology. I'm very glad I'm in cable TV."

Many college and university placement centers offer free or nominally priced occupational testing and counseling to alumni. A wide range of community colleges, state employment offices, and nonprofit organizations provide similar services to the general public, free or for nominal fees. Be very wary of "career management" firms that have been known to charge disproportionately high fees.

2. *Learn the Basics: The Letter and Résumé.* Particularly important is learning to write an effective résumé and cover letter. They are your calling cards. However, it is the cover letter, not the résumé, that should be the attention-grabber. When you consider that there are a hundred graduates with communications degrees for every broadcast or cable job available, it stands to reason that most résumés are created equal.

Your résumé should be a straightforward chronology of your experience and professional accomplishments, if you have any, packaged in no more than two pages. Make it concise and avoid smoke-blowing, which rarely fools anyone. It should be clear and tailored to the job specifications. If you're not sure how a résumé should look, use a professional résumé-writing service or check out such how-to books as Robert Hochheiser's *Throw Away Your Résumé,* Richard Lathrop's *Who's Hiring Who?,* and Melvin R. Thompson's *Why Should I Hire You?*.

The cover letter should give an indication of who you are, while the résumé blankets what you've done. A cover letter should be addressed to a person, not a title, such as Dear Mr. Vice President of Personnel. It should encapsulate your strong points in a few hard-hitting paragraphs. It is a *sales tool*. The cover letter should be to the point, while capturing in a choice phrase or two why it is you want to get into cable and some of the talents you have to offer.

From Dwight Tierney, of Warner Amex: "The thing that annoys the devil out of me is when my name is typed into a form letter. That person—and I don't care who he or she is—that person's résumé and letter go into the secondary pile. On the other hand, I recently received a package—a little box—at my office, and I signed for it and opened it up. Inside was a walnut shell with my name typed on a piece of paper around it. I peeled it open, then opened the shell, and there was a tiny piece of paper which I had to uncrumple. The message was: 'Dear Mr. Tierney: In a nutshell here's my backround. . . .' That person showed tenacity and creativity, and he was called in for an interview. Unfortunately, he wasn't as good as his walnut, but he got in the door."

Comments William Shaw, vice president of personnel at Turner Broadcasting:

"One of the common questions I hear from college graduates and people looking for jobs is: 'How do I get noticed?' The competition is stiff and a lot of people do have the same credentials. The résumé, then, is one of my pet peeves. How do I really know who this person is? There's always the person who writes in and says he attended Columbia. What does that mean—attended? Usually, it's a euphemism for 'dropped out.' It's better to be more up-front and write: 'attended Columbia for two years,' rather than trying to sneak one by. Another way I can tell who a person really is is by the way the résumé gets to me. Once, I got a box with gift wrapping around it; I opened it up, and it was a résumé. Another time, I received a summons that said: 'Johnny Smith versus Ted Turner. In lieu of $500 bail, you are to interview Mr. Smith, who wants to work in cable TV.' He got the interview, because it showed him to be a real thinker."

Turner Broadcasting Company is one of the largest employers in the cable industry; 850 people work at Cable News Network and another 200 are on the payroll at WTBS. Shaw continues:

"I can bet you that a good percentage of those people sent a cover letter with their résumé, too. You'd be surprised at how many people don't send a cover letter, though. Sometimes, I get a flood of résumés that look to me like they were sent to everybody in the industry. Obviously, it's recommended that you explore the job market in cable in its entirety. But the personal touch in a cover letter helps. It make take a little more time to prepare, but it makes me feel like I'm hearing from the person instead of a Xerox machine. It shows that you have a real interest in breaking in here."

Can a well-prepared and thought-out cover letter get you a job in cable TV?

Says Loreen Arbus, vice president of programming at Cable Health Network:

"The reason I hired the last person that I did was a wonderful cover letter. I've received over 1,900 résumés and cover letters, and I grade each one. Then they are filed under production executive or production assistant, so we can refer to them if a particular opening comes up. This one person, though, I couldn't file away. He wrote a letter that showed he had spent a great deal of time thinking about our mission at the Cable Health Network. It showed a great deal of research. There was a super amount of enthusiasm in the letter; it was warm and it was flawless as far as spelling and punctuation. I'm compulsive about that subject. I hate sloppy letters. If people can't take the time to look up how my name is spelled, how are they going to be sensitive to other people's needs? I can't believe the amount of letters I get that aren't proof-read at all.

"What is it that will send you over the edge? How can you differentiate yourself from 1,900 résumés? Take the time to write a letter that isn't a form letter. Write a letter that hasn't been sent to twenty other companies, while you hope desperately that someone will call. Write a letter that says you've taken the time to research what the company is about—and what the person you're writing to is about.

"I see a lot of letters that say, 'I'll call next week to set up a meeting.' Or, 'I would like to come in and find out what you're doing.' Well, I don't have the time or the interest to share what *I'm* doing. I want to know what *you're* going to contribute."

3. The Follow-up Phone Call. A number of human resource pros, and executives in the cable industry, advise that a prospective candidate call the targeted employer and cable company a couple of weeks *after* the cover letter/résumé has been mailed. At the very least, it will provide the job hunter with an updated feel for the employment possibilities at the particular company. At the most, it will give the job hirer a sense that you are really quite serious about a career in cable TV.

Mort Marcus, director of pay-TV sales for the Samuel Goldwyn Company, says:

"I get about ten résumés a week, and hardly anybody ever follows through. Some people just send résumés for a living, I guess. How can they expect a job? The person who follows through with a phone call is probably going to get more attention. Obviously, it's nice to know someone who knows me when you call. If I'm told that somebody I know told you to call, I feel like I have to take the call. But if you live in Oshkosh, Indiana, and you've never been out of there, and you've never known anybody who's left, you should still follow up the cover letter and résumé you sent with a call.

"I'll keep putting you off, I won't take the call, but if you're persistent I *might* take the call. By persistent, I don't mean phoning every day. There's a fine line between being persistent and obnoxious. I mean that you should try calling every three days or so. Finally, my secretary will say, 'Mort, it's so-and-so again, will you please take the call?' And I will. And I might be perturbed or short when I come on the phone, but you've got my attention. You also have twenty seconds, during which you can accomplish quite a bit. First, be nice and cogent. Get to the point right away. 'Hey, I'm going to be in Los Angeles next week and I have a couple of interviews set up at other companies. You may not have anything for me now, but I'm looking for fifteen minutes of your time. I just want to let you know what I can do.'

"Well, two things can happen after that. I may not have an opening, but I *will* remember your name. Or I might say: 'Listen, why don't you call so-and-so? He might have an opening now.' Or I'll say, 'Yes, you can give me a call when you're in L.A.' It's hard to say no when they ask for fifteen minutes of your time. You have to be a real ass if you say no."

Adds William Shaw (Turner Broadcasting):

"When it comes to the phone call, you have to be careful. You have to be up-front with the secretary about why you're calling. Once, I got a phone call from someone who told my secretary that it was a emergency, personal call. When I got on the phone, the caller said that he was looking for a job at CNN. 'And I really need the job, too,' he told me. From his standpoint, that was an emergency call and it *was* very personal to him. It was a ploy that was more annoying than creative. He didn't get an interview."

Another slice of advice: Don't call collect.

4. *Think of How You Can Help.* The password is differentiate. You must separate yourself from the other job-seekers, and show your potential employer that you have something to offer—something tangible and real. Show yourself to be a thinker. Ideas open doors. From Jim Ruybal, vice president of human resources, Daniels and Associates (an MSO that operates 26 cable systems and employs 700 people): "You have to be more innovative than the other 3,000 people who have a degree in their hands. I've been in cable now for three years, and I've told 200 people who wanted jobs that they should think of ways that they could help the industry and help the system. 'Come up with an idea,' I told them. If you do some research and give me an idea in marketing, I'll give it to

someone in marketing and that's one more person who will see your résumé and know your name. You know what? None of the 200 have taken my advice. I don't think any of them thought it would help, but it would."

Ruybal continues: "Be professional and be creative. If you're on the job hunt, go to a cable system and spend some time in the lobby. Hear the customers speak. Everybody's concerned with churn in the industry, so think of ideas how you would reduce it. You need intuition and innovation. When you've got an idea or two about programming or marketing or sales, call the system and ask for the department head. Take him to lunch! If he can't, take him to breakfast, or find out when he gets in to work and ask if you can meet him at his convenience in the parking lot for ten minutes. Would I meet with someone who said he had a good idea and wanted ten minutes? Damn betcha! I wouldn't say no, and if you're sincere but aggressive—without being obnoxious—and you follow it up with a three or four-page proposal with a couple of ideas, you'll be noticed. You'll become a face and a mind and not just a résumé. I mean, two of your three ideas might be terrible. You might be as green as gourd! But if you have one good idea—and one that shows you understand the problems, and you might be able to help—then you'll probably get a call back from someone at the cable system saying, 'Hey, maybe we ought to talk again.' "

5. *The Interview.* According to Kenneth N. Wexley, a professor at Michigan State University who is regarded as one of the nation's leading specialists in the human dynamics that occur during the job interview, it is critical to make a positive first impression. "The first four minutes is the time to put your best foot forward," he advises. Other hints include letting the interviewer do most of the talking, seeking to develop common bonds with the interviewer, and emphasizing something new and up to date about yourself. "Best of all," adds Wexley, "is to arm yourself with some fresh career-related accomplishment such as completing a course in tax-law changes or in small computer systems to tout in the interview."

One way to succeed at job interviews is to prepare yourself. Paul Shay of Showtime comments:

"A search firm called me and asked if I would be interested in the Showtime job. I said yes, but before I went into the interview, I wanted to learn everything about the company that I could. I investigated the company by going to the library, digging up the periodical index, reading all the articles and about the recent developments. If I interview someone here and they don't know about the recent developments in this company, it tells me they haven't prepared themselves. You have to be smart about the company you want to work for. You better know something about this company —and cable TV—if you come to interview with me."

And William Shaw adds:

"You have to be articulate on a job interview. If you come in here for an interview and I ask you a question, you shouldn't start mumbling and bumbling and end up with lockjaw." Personality counts. Also: position yourself as a professional. Never—ever—tell an interviewer that you can do anything if you only had an idea of what there was to do. Focus in on your skills—but be confident."

From Kay Koplovitz, president of the USA Cable Network:

"If your intent is to be a professional in the cable industry, then show that you're professionally minded. I don't advocate the old axiom that you should get your foot in the employer's door by offering to do anything they want. I certainly don't recommend for women to start out as gophers or in the mailroom if you want a management-level job. I don't like the 'I'll do anything' approach. I think it is important to fight and poke and search out the *meaningful* entry-level jobs. You have to know where the needs are and you have to believe you're the best person for the job. I got where I am because I'm the best person for my job. I'm more valuable than anyone else doing the job. And that's the way you have to think. You have to have very high expectations of yourself."

Paul Shay adds:

"There's a guy named Edgar Shied who talked about the three kinds of individuals that are in an organization. There is the conservative, who goes along with everything everybody says and doesn't bring much to the party. Then there's the radical, who is adverse to everything. Then there's the creative individualist—the one who brings something different but is not *completely* different. That person has the ability to identify something different and bring it to bear. When it comes to job interviews, I'm not looking for clones. I'm not looking for yes men. We want creative individualists. You have to establish yourself as a valuable individual."

And when it comes to the job interview, you had better dress for the occasion. Continues Shay:

"We don't want people punched out of a cookie mold, but if a kid walks in here with long hair and blue jeans, he probably won't make it. That's too radical. One thing I was taught a long time ago: If you want to move ahead, dress like your boss. If you don't have a boss yet, make sure your clothes don't stand out."

From Barry Black, the director of human resources and administration at ESPN:

"In the interview, I look for the person who can absorb what I'm saying. The ability to listen is a key factor. I also want someone who can articulate his or her career aspirations. There's a section in my interviews when I ask: 'What questions do you have for me? Tell me what's on your mind.' More often than not, I hear that the applicant has been knocking his head against the wall. Often, they say, 'What do I have to do to get into this business?' I tell them to go buy a football helmet. The point is, you can't let frustration and rejection get to you, particularly when you're at a job interview. If I ask why the person is here for an interview, if I ask what's on their mind, I want to hear, 'I have a basic fascination and love for the editorial focus of ESPN and the business of television. I realize that I'm not equipped at this point to move into a position that will afford me all I want out of life, but I want to put my nose to the grindstone. I want a chance.'

"When it comes down to it, a big turn-off for me is a young person who walks in here and gives you the opinion that he or she has tremendous industry savvy. I can see through that in a second.

It's far better to say, 'Through hard work and effort, and not thin air, I will become an asset to your organization. I want to *grow* with the company.' That's the kind of person who deserves consideration."

From Warner Amex's Dwight Tierney: "I interview all the interns we use at MTV and I have to get a sense of the person by the way they respond to questions. I'm not looking for the perfect answer, I'm looking for the immediacy, the positiveness. I get a lot of people who come in with degrees in communication—and who specialize in production—and I ask them, What can you give me? And what can you get from it? I also ask, What have you done to plan to get where you want to be? I get a lot of blank looks, and I want a definitive answer. When it comes down to it, I want a kid who has been thinking about cable television. I want someone who is willing to work as a gofer, pull some cable and lug the mini-cam, and I want him or her to say, 'I'm going to show you how good I can be. I'll go from a gofer to an assistant producer to a producer and then into programming. And I'm willing to work overtime for free.' In other words, you have to be willing to get your hands dirty. To use an old football term, I want someone who's willing to 'give up their body' to the cable industry—and Warner Amex."

And from Juanita Desosa, personnel supervisor at the Valley Cable system in Southern California, "During the job interview, it's very important for the applicant to look me straight in the eye. It sounds real easy, but a lot of people don't know how to do it. *Learn* how to do it. If you can't look me in the eye, it tells me that maybe you're not ready, or don't have the inner confidence, to make it in cable TV."

6. *Networking.* Turner's William Shaw: "If you want to get into cable, I'd try like hell to meet people who work in it. I'd go the local church or to a bar, or any place you can make a contact."

Perhaps the most important skill to master is the skill of "networking"—the process of using "who you know" contacts to sniff out job openings, obtain key referrals, and make connections with people capable of hiring you. Show business—and that includes cable television—is the most incestuous industry around. People get hired through the grapevine.

If you think you don't know anyone, think again. One career planner suggests starting out by taking a blank sheet of paper and listing all the people you have known since high school—relatives, friends, college roomates, teachers, and so one. People are invariably surprised at how extensive the list soon becomes. The next step is to zero in on those people who are likely to be best placed to hear about job leads in the cable industry, and begin calling them up one by one. A friend in advertising, for example, might very well know some people in cable.

Another way to "network" yourself into a job is to get on the mailing lists of cable organizations, such as Women in Cable (which has twenty-five local chapters), or the NFLCP and the NCTA, both in Washington, D.C.. Through their newsletters and press releases you will learn of seminars, workshops, and conferences offered. Attend as many as you can. Mingle.

Janice Thomas, advertising director of the Black Entertainment Television (BET) cable network, comments: "If you know a janitor who used to scrub the floors at a cable company, talk to him. You have got to give yourself the chance to meet as many people as possible. And follow through on the leads."

In order to mobilize your contact network, it is necessary to keep your eyes and ears open at all times. You never know when your "who-you-know" is going to walk right through the door. Says Liz Carney, affiliate sales representatives for the Spotlight cable network:

"I was going to Northeastern University in Boston, and was working part time as a waitress in a place called Friar Tuck's Pub downtown. My major was communications and journalism, and I was looking for a job in television. One day, one of the workers from the pub called me over and told me that the people I was serving lunch to worked in cable TV. So I went up to them, asked if everything was okay, and then mentioned that I knew they were in cable and that I was interested in getting in. One of them mentioned their company name, and told me to come into the office to see if there was anything available.

"Before I went down, I went to the Massachusetts Cable TV Commission in order to find out some things about their company. While I was there, going through a few publications, I mentioned what I was aiming for to someone who worked at the Commission. Then I opened my mouth again and asked if they needed any help. The answer was yes, and I was told that, if I was just starting out, the Cable Commission was a good place to get a broad view of the industry instead of gophering at a cable company. So I got my first job in cable at the Massachusetts Cable Commission. And it all traced back to the guy who ate lunch at the pub. You know what? He left me a pretty good tip, too."

7. *Do Your Homework.* The best place to begin preparing for a job in cable television may well be the research area at your local library. Paul Shay talks about going to the library and pouring over old *Wall Street Journals* and the *Cable Data* volume to find out more about Showtime. Another book that lists every cable system in America by state is the *Television Factbook,* put out by the *TV Digest* people in Washington, D.C. It will help you find out which systems are in your area, and which ones you might want to check for possible jobs.

The trades are a wonderful way to keep up to date. *The Hollywood Reporter* recently ran a piece about the new Sports Network, cable's first multi-regional pay sports service, and mentioned that owner Group W planned to hire 500 people for this venture. Those kinds of clues emerge every week in the Hollywood-based trades. Weekly *Variety* has a cable section and usually can be found in local libraries. *Broadcasting Magazine* can be a good source, as well. The cable trade publications—*Cablevision, The Multi-Channel News, Cable Business,* etc.—all contain articles that will help you prepare for your interview or plan of attack. If there is a specific cable company that you have your sights on, look up the company annual report, consult business and financial directories, and perhaps call up the company to ask for any material that might be available. I know one person who called up a cable network and told the receptionist he was doing a story on cable for his college paper. He received a ton of material from the network's press department, which he used to prepare himself for an interview at the company. The kicker was that he had been out of college for two years.

The key point here—and it can't be stressed enough—is that no company will hire you unless you have a solid inkling of what the company does to pay its rent. If you're interested in working at the Playboy Channel, for example, it's not enough to know that Hugh Hefner smokes a pipe.

From Lynn Edelstein, director of personnel for Playboy Enterprises on the West Coast, which includes the Playboy Channel: "I have a question that I ask in my job interviews: How did you find out about the Playboy Channel? The answers vary, but it's a good sign if the person has read a good deal about it. You really have to find out which direction the company is going and its general profile. Yu have to sound knowledgeable. If someone walks in here and asks me what's Hef really like, that person probably won't get the job at the Playboy Channel. You have to ask questions and have answers that show you have a good idea of what the channel is all about. You have to show that you've invested some time."

For possible job opportunities, keep your eye on the classified ads in the back of the trade publications. Also stay on top of the Help Wanted sections of the local newspapers, particularly in Sunday editions. A recruitment ad will often pop up from a cable concern. Besides library research and the classifieds, you can gain valuable inside information—along with networking contacts and free publications—by attending association conventions, trade shows, cable seminars, luncheons, and so on.

A telling thought from Dwight Tierney, of Warner Amex: "There's only one thing that really galls me, and that's when a job seeker doesn't know what we do."

8. *Timing.* A little bit of luck won't hurt. Sometimes, it's as easy as walking through the door on a day someone has just passed you

on the way out. Of course, the better prepared you are for the job —and the hunt—the better chance you have to turn good fortune into a fortune—or a weekly paycheck from cable TV.

Michael Fuchs, president, HBO Entertainment: "There are people hired here sometimes who come in the day that someone decided to get married and move to British Columbia, or someone got fired, or someone got transferred, and, all of a sudden, there's a job opening. If you've got the right credentials and the right attitude, we tell you that you've got the job. If that person had come in a day later, the job might have been filled. So it isn't just qualifications. It can be a matter of luck. I've got to believe there are a good many people on the outside who have the ability to work inside HBO. Sometimes, all it takes is a stroke of good timing." For a good story about great timing, see the profile of MTV's Martha Quinn, page 148.

9. Determination. Job hunting contains many facets of a crapshoot; the process can become protracted and painful. In fact, studies have suggested that on a list of life's most stressful experiences, job search ranks number two for many people—just behind the loss of a loved one. However, job hunters are fueled by the knowledge that cable television's growth will continue and job opportunities will increase.

William Shaw of Turner Broadcasting:

"You have to be resilient. You have to keep your chin up and knock on doors and you can't be put off by rejection. You have to fight for it. You have to get *Turnerized.* You have to be full of piss and vinegar. People in cable want to hire the people with the fire in their eyes. You have to be a self-starter and you have to have the grit. No one wants to hire a passer-by. No one likes a yawner. It's important to go out there and turn it on!"

Harlan Kleiman, an independent cable producer:

"You have to have a unique presence. The hardest thing for a young producer to understand—or a young person looking for a job—is that you may have to stick your finger in the phone thirty or forty times. You may have to pummel and pummel and pummel. You may have to find every angle to get in. Getting in the first time is a lot harder than getting in the second, unless you were really horrible the first time.

"You've got to keep on pushing until you get there. You have to orchestrate. And the key thing is hanging in. You *have* to have perserverance. And you have to want that job more than you've ever wanted anything in your life."

And finally, from Barry Black of ESPN:

"You have to withstand the negative to get to the positive. You have to say, 'How do I get from here to there?' and you have to stick to your game plan. You have to persist. That's the most important part of finding job in cable TV. You can't get down on yourself. If you don't find something right away, you have to pick yourself up, dust yourself off. You can't let rejection get to you. You always have to remember: there are a lot of reasons why people don't get jobs. It's not only you. You also should *learn* from every interview and every day you're looking. Keep at it, and keep absorbing.

"I know a number of people who work at ESPN, and throughout the cable industry, who had that resiliency. To them, finding a job was like riding a bicycle. If you fall off—if you don't get a job right away—get right back on the seat. Concentrate on the road ahead . . . and *steer.*"

NINE

The Future in Cable

For cable operators—indeed, for anyone who wants a career in cable—the 1980s promise to be the best of times and the worst of times. There's no question, of course, that the industry will continue to grow at a healthy rate between now and 1990, and some applications of cable technology that are now strictly experimental will be commonplace by the end of the decade. But how fast will that growth take place? Which of the cable networks will survive and which will fail? Which interactive applications of cable technology will catch the public eye and which will flop? And who stand to profit in the long run? The answer to any of these questions, it seems, is anybody's educated guess.

One bullish forecaster is Gustave Hauser, former chairman of Warner Amex Cable and now a cable consultant. By 1990, Hauser predicts, 80 percent of the nation's 100 million television homes will be wired for cable. The average subscriber will also pay for two or three premium channels and a scattering of pay-per-view events, he believes, while fully half will be hooked up via cable to fire and burglar alarm systems. And, Hauser predicts, as many as one in eight homes will be equipped with a home computer connected to a cable company's two-way lines for information exchange, shopping, and banking from the privacy of their own living rooms.

Hauser's enthusiasm is shared by other optimists. Richard

Munro, president of Time Inc., a high roller in cable circles, paints a rosy picture of cable's future: "The wired highway across America, so long heralded and now a reality, will have a hundred lanes: some fast, others slow, some a hard pull uphill for serious viewers, others a razzle-dazzle downhill slide for entertainment fanciers. . . . We will witness the birth of new kinds of entertainment, new art forms, abstract entertainments. . . ." Joe Bain, president of Cablevision Systems, another industry heavyweight, is looking forward to the coming "programming boom." If an individual operator can't sell 90 to 95 percent of his potential subscribers, Bain argues, "then someone is not doing his job."

One prestigious consulting firm, International Resources Development (IRD), has even gone so far as to put a price tag on what consumers will likely be willing to pay for this cornucopia of cable riches. In addition to fees for basic cable, premium services, and $10 or more a month for pay-per-view events, IRD predicts that subscribers will shell out a staggering $78 a month for such interactive services as electronic newspapers and weather forecasts, video games, shop-at-home channels, and two-way banking.

Not all analysts share these lofty sentiments, however. A 1982 survey of Wall Street media experts produced a consensus that the cable industry would continue to grow, but estimates of how many homes would be wired for cable in 1990 ranged from 43 million to 55 million. The analysts cited high capital costs of construction, increased regulations, and competition from noncable delivery systems as the biggest stumbling blocks in the field.

Other voices of moderation find no fault with the projections of big dollars and big profits throughout the 1980s, but they wonder who's going to reap those profits. The capital investment required can be tremendous, and many smaller systems may find themselves devoured by giant MSOs just to remain competitive. There's also a threat that Bell Telephone, for a long time legally banned from entering the information systems market, will aggressively pursue that market in the 1980s. Ma Bell already has a line into virtually every home in America, unlike cable operators, who still haven't entered some of the richest big-city franchises in the country.

The explosion of programming "for cable only" also worries some experts. Ted Turner, whose Cable News Network is usually cited as cable's biggest success, has said: "Too many services are being launched too quickly to survive." On a typical cable system, most viewers will continue to watch the three commercial networks and one or two major premium services, leaving all the "cable-only" competition to fight for a small fraction of the total audience, hardly enough to attract strong major advertising support. Sean McCarthy, director of development for Time Inc.'s Video Group, doesn't think many of the independent channels will turn a profit. "I believe they will begin dropping off around 1985," he predicts, "and only the strong will survive."

The experts may have some trouble charting just how fast the cable business will grow, but there's no question that it will continue to grow at a healthy rate and to experiment with new technologies at an increasing clip. The big story for cable television in the 1980s will be how these new services will translate into jobs.

Even if none of the new technologies takes off, employment prospects for the cable industry are rosy. Between 1975 and 1981, according to the NCTA, the number of salaried positions in the industry rose better than 300 percent, from 12,727 to more than 51,000, with the current workforce numbering close to 70,000. And that figure doesn't even include companies that provide programming and manufacture equipment for cable operators. Possibilities for rapid advancement are excellent in the cable business, too: at Cable News Network, for example, a talented trainee can move from a minimum-wage, entry-level position to director or producer within one year. "Look at those faces," said Julian Goldberg of MTV, on the age of his programming staff. "At the broadcast networks, some of them would be 'gofers.' Here, they're directors, producers; they're in charge." Similarly, an installer in a brand-new system can find himself supervising a crew of thirty workers in a short time if the territory lives up to its growth potential.

Of course, as the entertainment, information, and interactive branches of the cable industry join forces and grow, new skills are going to be in sharp demand. To get a sharper picture of what those skills are likely to be, let's look at some of the alternate futures on the cable horizon.

THE CABLE SYSTEM

Whether your local cable operator is a Ma and Pa entrepreneur or an arm of some huge MSO, chances are he has job openings. And until those operators begin to approach the limits of their systems' potential, there will be a continued need for both skilled and unskilled workers.

On the simplest level, the firms involved in every arm of the cable industry are like any other—they need secretaries, clerks, accountants, and other general office help. And since operators string more than 14,000 miles of coaxial cable every year, they'll need workers to run the cable from pole to pole, or, as some communities demand, to bury it underground. Still more workers will have the job of tapping those feeder lines and running individual drops into customers' homes.

Once the business structure and the cable network are established, someone has to keep the system running from day to day. That calls for system planners and engineers at the headend, along with a fleet of troubleshooting technicians in the field. The increasing use of computerized switching and delivery equipment often means a need for computer programmers as well.

The biggest demand for employees in a new consumer-driven industry like cable TV comes from the arm of the industry that brings in the revenues from the public: sales and marketing. Door-to-door soliciting and telephone sales are the most common jobs here. There is also a need for customer service personnel to take complaints and soothe irate subscribers. Usually, an ability to get along with the public is more valuable than experience in these positions.

There are many openings for anyone with a BA or more advanced college degree. Besides company management, there are slots for attorneys and engineers to prepare complicated franchise applications and assemble technical reports. One sleeper job classification could be that of market researcher: most big corporations don't feel comfortable investing the huge sums of money required to play the cable game until they've studied the enterprise from every single angle.

That's not necessarily all, either. New needs crop up with new systems, and the big MSOs with their sophisticated interactive systems promise to be on the leading edge here. June Travis, senior vice president of American Television and Communications (ATC), the largest MSO in the country, admits that "the business is changing on us. The industry is opening up whole new fields and systems. New jobs are opening up all the time in every area." Adds John Dawson, ATC's vice president of human resources, "Cable, historically, has been a fairly insular business. It's clear we can't be that any more. We need people who can speak the language of business, and we need them to have good interpersonal skills. Cable television today is a giant mosaic. There are numerous opportunities, particularly in the technical side and in marketing and sales, for people who have a grasp of what this industry is—and where it's going."

ENTERTAINMENT PROGRAMMING

There's one inescapable fact about cable systems: Their appetite for programming is tremendous—and likely to grow even more voracious in the next few years. These days, no self-respecting MSO would think of venturing into a new territory with anything less than 50 channels, and some boast a capacity of close to 100! Increasingly, too, operators of primitive twelve-channel systems are finding it economically essential to upgrade to thirty-six or more channels. And since nature and cable operators both abhor a vacuum, they'll be looking for programming to fill all those slots on the dial.

We'll get to some of the nontraditional forms of programming like videotex and interactive services in just a minute. Those are interesting experiments; but as long as television is part of cable TV, the biggest chunk of channel space will go to entertainment —including news, sports, weather, and movies. Despite the traditional look, however, they're likely to come in some surprising packages.

One approach to winning a cable channel is to start your own network. That's what USA and ESPN have attempted, with modest if less than spectacular success. That's also the theory behind "superstations" like WTBS in Atlanta and WOR in New York. But industry analysts are pessimistic about the chances of any cable service, outside of HBO, ever making a serious run at the established TV networks, which is why ESPN, for example, deals exclusively with sports. Similarly, USA tends to program in blocks— evenings and weekends are for sports, daytime hours are for women's programming, Friday and Saturday nights are set aside for rock 'n' roll, and so on.

It's called the "narrowcasting" approach. At last count, more than two dozen companies had targeted specific audiences and launched cable channels directed specifically to those audiences. For news buffs, there are two versions of the Cable News Network and also Satellite News Channel. The ARTS and Bravo Channels are aimed at culture-loving, upscale audiences. The Disney Channel hopes to carve out a healthy chunk of the family audience, while Playboy's titillating service is no doubt aimed at an entirely different group of viewers. Country-music fans have the Nashville Network. Minority programming is served up by Black Entertainment Television (BET), the Spanish-language Galavision, and a smattering of satellite-delivered signals for Iranians, Jews, and

Japanese-speaking viewers. There's MTV, with twenty-four-hour rock music, the Weather Channel, and a slew of services aimed at fundamentalist Christians.

With that many players, there will certainly be no shortage of opportunities in the cable programming business. With the proper combination of enthusiasm, experience, and good luck, it should be possible to land a job at one of the many programming services —on either side of the camera. Don't expect too much job security, however, because turnover in the industry is likely to be tremendous.

The problem with narrowcasting is that it requires advertiser support. Advertisers, traditionally, are unwilling to part with hard-earned dollars unless their message is going to a large audience— something that narrowcasting deliberately eschews. Time Inc.'s Sean McCarthy argues that "it is economically very difficult for any channel to get by with under 5 percent of the viewers." In the TV game, with the three networks and the major pay services like HBO and Showtime accounting for an estimated 80 percent of the available viewers, competition will be fierce for the remaining 20 percent. Already, the highly regarded and well-backed CBS Cable, which took the high road in its programming, has gone under.

Some services, of course, don't depend on advertising to pay the bills. Most of them are premium movie channels, and while it doesn't appear that any are in immediate trouble, there also doesn't seem to be much room for new faces. The notion of paying $10 to $50 a month for the right to see uncut movies is sound, all right, but cable moguls have discovered that there is a limit to the total bill even the most well heeled subscriber will pay. All of this means that the rich—namely, HBO and Showtime (which recently merged with the Movie Channel)—will continue to get richer over the next few years, while the competitors will, at best, sweat it out.

One area of premium television that is only now beginning to be explored is the notion of subscription sports. In Chicago, for example, you can only see certain White Sox games if you subscribe to the right cable system or to a microwave-delivered version of the same signal. Other premium services have collared the rights to selected home games of popular teams like the Los Angeles Lakers. If you want to see the game, you'll pay for the privilege.

But the real money to be made in programming in the years ahead, say veteran industry-watchers, is in pay-per-view (PPV) programming, especially sports and special events. Former Warner Amex chief Gustave Hauser enthuses: "The growth of (pay-per-view) will make it the world's largest movie theater." Cable analyst Paul Kagan calls PPV "a sleeping giant" that could account for yearly revenues of $2 billion by 1990.

Already, PPV has scored some stunning successes. The September 1981 championship fight between Sugar Ray Leonard and Thomas Hearns, for example, pulled in $7 million from cable and subscription TV (STV) viewers, who paid between $10 and $20 for the privilege of watching what turned out to be a very good fight. Reaction was also positive to a showing of *Star Wars* in 1982, before the film began playing on premium services like HBO and Showtime. Nearly 325,000 viewers paid an average of $8 apiece to follow the adventures of Luke Skywalker for the umpteenth time. The latest trend is to premiere a movie on cable the same night it opens in theaters, beginning with *The Pirates of Penzance*.

There have also been some less-than-successful PPV offerings. In 1982, a live broadcast of the hit Broadway musical *Sophisticated Ladies* attracted a meager 60,000 viewer for the STV and cable operators offering the show. Rock concerts by the Rolling Stones and the Who were more successful, but they also didn't attract the blockbuster numbers that have come to be associated with pay-per-view hits.

What's slowing the growth of PPV is the snaillike pace of technology. In order for a cable home to be equipped to pay for programs one by one, a so-called "addressable" box has to be attached to the television set. Cable operators have experienced technical problems with the boxes, and many have been reluctant to spend the money to convert their systems to accommodate pay per view. Only about 500,000 homes scattered around the country are currently equipped for pay per view. That number, however, is expected to swell rapidly. "Five years from now, it could be 20 million," said Gustave Hauser. By 1986, Group W officials expect to make pay-per-view service available to half of their projected 4.5 million subscribers. Once that day comes, PPV rates for run-of-the-mill events could be pegged at as little as $1. But there's a real chance that you'll have to cough up $10 or more to watch the Super Bowl and other major events that are now received "free." How big a profit center can PPV be? Paul Kagan cites as one possibility the potential of the 1988 Olympics: For an "unlimited pass" to the two-week games, he projects a revenue of $1 billion, without a single commercial!

At worst, PPV is sure to pay for the investment in two-way technology that it will require, which opens the door for a multitude of new services that will dramatically alter the face of cable TV. With two-way addressable convertors in wide use, such speculative cable uses as opinion polling, electronic mail delivery, energy management, video shopping and electronic banking, videotex, and even sophisticated security systems become possible. Alone, none of these has the profit potential to justify the huge cost in convertors and additional cable required; PPV provides that financial justification. John Messerschmidt of Scientific-Atlanta, a leading manufacturer of cable decoding equipment, explains: "You need one foundation service that makes two-way attractive financially. . . . Two-way for premium programming will be the foundation service."

VIDEOTEX AND TELETEXT

For years, one of the biggest differences between cable and broadcast TV has been the inclusion of channels devoted exclusively to alphanumeric data. Depending on the individual system, this data can take the form of news headlines, weather forecasts, "video classified ads," program listings, or anything else that lends itself to translation into words. Teletext and videotex are similar in that both use alphanumeric displays, but the differences between them are profound and promise to become even more pronounced as technology improves.

Teletext is the simplest form of interactive data use available to cable subscribers. Through a process known as "vertical interval blanking," data is transmitted over an unused portion of an existing channel frequency. Using an inexpensive ($50 is about average) decoder, the cable-user calls up a specific "page" from the continuous cycle of data; the decoder stores that page in memory the next time it appears in the cycle, and the page is then displayed on the television screen until the user cancels the request.

Videotex is a far more complicated system with greatly increased memory and far more flexibility; however, it requires either a telephone hook-up or, ideally, a two-way cable system. Vast sums of information are stored in a central computer memory at or connected to the cable system's headend. To call up a specific "page" of data, the user consults an index, then punches in the appropriate request on a computer terminal or a decoder (average cost: $300 and up). Unlike teletext systems, which cycle through the available data continuously, videotex systems are accessible virtually immediately.

What kind of information is suitable for a teletext or videotex system? The possibilities are endless: airline schedules, restaurant listings, newspapers, classified ads, the Yellow Pages, encyclopedias, sports results, and stock prices are just a few. Great Britain's state-run Prestel service was one of the first videotex ventures to take to the airwaves, with only limited success. Meanwhile, in the United States, a score of cable and information specialists have launched videotex and teletext experiments in preparation for what everyone expects will be the next wave of cable technology.

The first serious videotex experiment on these shores was launched in 1980 by Dow Jones Cable Information Services. Today, Dow Jones's package offers a combination of news, weather, sports, movie reviews, and an encyclopedia. Other pilot videotex and teletext projects currently under way are being sponsored by, Knight-Ridder Newspapers, Times-Mirror Publishing (including the Los Angeles *Times* and *Newsday*), and American Bell Consumer Products, in association with such cable giants as, Cox Cable, Warner Amex, and Time, Inc.

When do the experiments end and full-scale services begin? Once again, the hang-up is technology, this time flavored with a dash of bureaucracy. Not only are two-way systems still a vision of the future, but in the case of videotex, there is no agreement among the major players as to which system will be adopted as standard. And no one wants to leap in feet first until he can be sure his system will be the standard.

Once the standard is adopted, however, videotex promises to be profitable simply because of the wide range of sources that will pay for either providing or tapping into the service. Industry experts see three different avenues through which text services will pay their way. First, there are the providers, such as airlines and restaurants, who could be asked to pay by the listing or by the page to have their services included in the data bank. Second, there is "pay teletext"—the user pays a flat monthly fee for unlimited use of the service. Field Communications (WFLD) in Chicago is working on plans for such a service, with a projected monthly charge of between $30 and $120. Finally, there's "pay per access," a system in which the user pays an agreed-upon charge for either time or amount of data used. Oak Communications is exploring this version of videotex in its pilot projects.

Teletext and videotex have one overwhelming attraction for cable system operators. As Martha Johnson, director of administration of the Videotex Industry Association, admits: "Cable operators may want to offer these services as a way to fill channel capacity." Of course, the movers and shakers in this growing field are convinced that the marriage between text and cable television will lead to more than just a space-filler.

TRANSACTIONAL SERVICES

John Carey, of New York University's Interactive Television Program, maintains that consumers disdain "a lot of *information* on videotex. What they want is the ability to conduct certain transactions—a little banking, a little shopping." And as has been the case throughout cable's history, what people want, they tend to get. In the case of video shopping, the first step, short of a sophisticated two-way system, is an LO program concentrating on services and products from specific manufacturers.

One such program has been available to cable subscribers in Peabody, Massachusetts, since early 1982. Developed by the J. Walter Thompson advertising agency and the Adams-Russell cable system, *CableShop* lets viewers phone in requests to see three- to seven-minute "infomercials" from local and national advertisers. Surprisingly, say the executives in charge of the program, most viewers simply tune in at random rather than requesting specific bits of video salesmanship. In San Jose, California, a cable operator has taken the concept a bit further with *TV Auction,* a weekly live broadcast from the San Jose Flea Market. Viewers phone in bids on merchandise ranging from pocket knives to microwave ovens. Reportedly, it's a runaway success. In Portland, a cable system offers "Barter Time," a thirty-minute program featuring the swapping of merchandise.

Other, more sophisticated programs around the country offer the equivalent of in-depth commercials promoting everything from home appliances to real estate. One of them, "The Cable Store," showcases merchandise for a national viewership via SPN. But experts think that video shopping will really take off when it's tied to the next logical cable service: video banking. Already, Chemical Bank in New York City has tied two hundred Manhattan households into an experimental network called Pronto. In Florida, Knight-Ridder and AT&T have put together a consortium of local banks that ultimately will form a full-scale electronic banking system.

Pilot shopping services include American Can's HomServ and Times-Mirror's Home Shopping Channel, which is tied into Comp-U-Card, a budding credit card company. The philosophy behind all these services is simple: viewers equipped with a computer terminal or the equivalent can call up catalogue listings of a specific category of products, select the one they think best suited to their needs, and arrange delivery by punching a button and having funds debited from their preestablished account. "Probably the biggest problem with cable shopping," says Time Inc.'s Gerald Levin, "is that the user can't squeeze the fruit."

OTHER INTERACTIVE SERVICES

When discussing the future of cable, media-watchers often point to Warner Amex's QUBE system, the first of which was built in Columbus, Ohio, and is now expanding to an estimated 1 million homes in Pittsburgh, Cincinnati, Houston, Dallas, and the suburbs St. Louis. QUBE is a sophisticated two-way system that offers a wide range of interactive services, including PPV. What has the cable experts most excited and concerned, though, are the home security and polling capabilities of QUBE and other two-way set-ups.

Home security promises to be perhaps the biggest thing ever to hit cable TV. As of mid-1982, more than one hundred cable systems nationwide either offered or planned to offer security as part of their package to subscribers. Mike Korodi, president of Warner Amex Security Systems, says flatly: "It's a multimillion-dollar business." Significantly, Korodi began his cable career with the Columbus QUBE system.

The home security option is really three services in one: a smoke/fire alarm; burglar alarm; and medical alert. The first step for the cable subscriber is to have someone, either the cable operator or an affiliated security company, install the necessary smoke and entry detectors along with a button to push for medical assistance. Once the elements of the system are hooked up to the cable, a group of computerized monitors continually scans the system. When a "live" emergency signal comes in, it's verified by a human operator, and the appropriate emergency personnel are sent to the customer's home.

Is there a demand for cable security? You bet there is. Despite a hefty average installation charge of about $1,200, more than 10 percent of Columbus's QUBE subscribers have already signed up: 7,000 homes out of a potential 58,000. Warner officials say their cable alarm system has resulted in 30 to 40 percent fewer false alarms than conventional systems.

The other, much-heralded QUBE service is instantaneous polling of viewers on questions ranging from the trivial ("What's your favorite brand of laundry soap?") to the profound ("Who is your choice for President?"). Supporters of this service call it interesting, although they're not sure whether it will be important or not. But detractors are worried about the Orwellian implications of all two-way technology: Janet Maslin of the *New York Times* wonders, "What if a serious political poll was being conducted and the buttons were being pressed by three-year-olds?" And another observer admits, "What you can do under the guise of home security is phenomenal."

The upshot of those difficult questions will probably be more work for lawyers, but not a slowdown in the growth of new cable technology. What kind of additional skills are likely to be of value to the interactive cable system—and the cable industry—of the future? For starters, there will be the usual assortment of openings for office personnel, executives, and engineers because despite their revolutionary potential, these services still closely resemble their traditional cable counterparts in terms of the need for manpower. "In addition, I think one of our greatest growth areas will be in people who can sit down at a terminal and understand what it is," said John Ford, the vice president of human resources at HBO. "I think the corporate focus is directed at the communications technology. I think there will be less concern with what's on the screen but how it gets there."

Probably most in demand will be workers with hybrid skills: part writer, part engineer, part computer programmer, and part wizard. Take videotex, for example. There's an art to formatting and indexing data for maximum effectiveness that demands something more than either a writer or a programmer alone can supply. In the realm of polling, there's also an art to writing questions objectively to elicit the most accurate answers.

Perhaps the greatest qualification any career-minded individual can possess, though, will be flexibility. "The people who tend to make it in this business are broad-gauged," says John Dawson of ATC. "The cable business is so new, and it's emerging so fast, that there's no such thing as a vertical pipeline to a job. For instance, we're putting in a pilot project of addressability in our San Diego system, and we need someone to ride herd on that. The job is not on any organizational chart. The person who is going to get it is the person who has a certain mobility and who understands that the various pieces of cable are interdependent."

Adds John Ford of HBO, "If you come into this company, you have to be prepared to go from X to Y to Z. You have to have broad-based talent. We need the information processors, sure, but there is less of a pyramid system. Most of all, we'll need self-starters."

Cable television executives paint a rose-colored future for the industry. Most predict that cable ultimately will be the cornerstone in people's home-entertainment centers. "We're going to be the richest and most powerful group of people that ever existed," concludes Ted Turner, who should know. "And there are a lot of opportunities to get in on the action. "Of course," he adds, "we can't have a nuclear war. That would wipe out just about *everything.*"

Cable-Chat: A Glossary of Cable Television Terms

If you intend to speak with the natives, on a job interview, for instance, it is imperative that you know a few words in their language. This knowledge will not only impress the native behind the desk but will help you understand what said native is grunting when he asks you a question. When it comes to the big job interview, don't forget what William Shaw, who hires at Turner Broadcasting, said: It's important that you relay the feeling that you comprehend the industry you want to join. A good way to impart this feeling is to toss in a few syllables—at appropriate times, of course—that only true-blue cablephiles can comprehend.

Hence, a quick Berlitz-type course in *cable-chat,* the language of the cable revolution.

Access Channel   Generally speaking, the channel(s) set aside by the cable operator for use by the public, educational institutions, local government, or for lease on a nondiscriminatory basis.

Ace   (Award for Cablecasting Excellence) The cable television industry's highest award for original, made-for-cable programming, both locally and nationally. The annual awards are sponsored by the NCTA (National Cable Television Association).

Aerial Plant   Cable that is suspended in the air on telephone or utility poles.

Alphanumeric Keyboard   A keyboard that allows communication with a computer in letters and numbers on the TV screen, rather than moving images.

Amplifier   A device that boosts the strength of an electronic signal. Amplifiers are placed at intervals throughout a cable system to keep signals picture-perfect.

Aural Cable   The origination of radio programming on an FM channel leased from an existing cable system, available to cable subscribers only. Also called CABLE RADIO.

Bi-directional Cable   A fancy way of describing "two-way" cable communications.

Bird   Cable slang for communications satellite, i.e., "HBO is up on the bird."

Broadband Communications System   Occasionally used as a synonym for cable television. It can describe any technology capable of delivering multiple channels and services.

Cable Television   A communications system in which television signals are transmitted by means of a coaxial cable wire and/or optical fiber to the TV receivers of subscribers, who pay a monthly fee for the privilege. The signals can be picked up directly from a television camera in a local studio or taken off the air from local TV stations; they can also come from distant locations via landlines, microwave, or communications satellites (birds).

CATV   (Community Antenna Television): The term first used to describe cable television, which got its start in the early 1950s providing television reception to remote communities via a tall antenna. The term continues to be used occasionally to refer to cable service, usually by the cable pioneers.

Channel Capacity: The maximum number of channels that a cable system can carry simultaneously. Channel capacity on an individual system can range from 12 in the older systems to 127 in the new-builds.

Character Generator (CG)   An electronic device that displays letters and numbers on the television screen.

Cherry Picking   What a cable operator does when he takes programs from different satellite services and combines them on one channel, rather than giving a channel over entirely to a single service.

Churn   An industry term that describes the canceling of cable service by a subscriber. A subscriber can describe to yank the entire cable sevice or just a pay channel, while keeping his basic cable intact. "Churn," of course, is a dirty word in cable TV. The term also describes what it does to a cable executive's stomach.

Coaxial Cable   The actual line of transmission that distributes TV signals to subscribers' TV sets in most cable systems. Its principal conductor is either a pure copper or copper-coated wire, covered by a layer of insulation, which is in turn encased by a tubular shielding of braided wire or an aluminum sheath that serves as the secondary conductor. An electromagnetic field is created between the inner wire and the shielding, which keeps the signals from being attracted off course by magnetic fields in the environment.

Communications Satellite   An orbiting space device that retransmits signals received from earth to other receiving points over a wide area. Since 1975, communications satellites have been used in the United States to distribute cable programming nationally.

Convertor   A device, set on top of the television set, that can increase the channel load of the set so as to accommodate the multiplicity of channels offered by cable television.

DBS   (Direct Broadcast Satellite) A proposed system in which TV signals would be transmitted directly from a satellite to receiving dishes located at individual homes. A potential threat, obviously, to cable television as we know it.

Dish   A parabolic antenna that receives TV signals from a satellite or sends them to it (also known as an earth station). The term "dish" is also used for an antenna that receives microwaves.

Distant Signals   A television channel from another city imported and carried locally by a cable television system.

Downlink   An earth station that receives TV signals from a satellite.

Downstream   The flow of signals in a cable system from the headend to the subscribers.

Drop Cable   The piece of cable that connects the TV receiver in a subscriber's home with the piece of cable—the feeder line—in the street.

Feeder Line   The cable that serves as an intermediate link between a system's main trunk line and the short lengths of cable, known as drops, that enter subscribers' homes.

Fiber Optics   The use of very thin and pliable tubes of glass or plastic to carry wide bands of frequencies. Currently, this technology is still in the experimental stage.

Footprint The territory within which a particular satellite's signals can be received.

Franchise The contractual agreement between a cable operator and a local governmental body that defines the rights and responsibilities of each in constructing and operating a cable system within a specified geographical area. Without a franchise, a cable company can't build the cable business. Franchises have been hotly pursued by numerous cable companies across the country.

Franchising Authority The governmental body responsible for specifying the terms of a franchise, awarding it, and regulating its operation. While the franchising authority is usually a local city or county body, some areas are regulated exclusively on the state level.

Headend The electronic control center of the cable system, at which TV signals are collected and sent on to subscribers. The headend is the site of the receiving antenna and the signal-processing equipment essential to the proper functioning of a cable system.

Home Box Office (HBO) The first pay-TV network of movies and special programs, which began service in November 1972. HBO is by far the largest pay service, with a subscriber count of over 12 million.

Homes Passed The total number of homes an individual cable system is able—or has the potential—to serve.

Independent System A cable television system individually owned and operated, not affiliated with an MSO (multiple system operator).

Interactive Cable A two-way system that has the capability of carrying signals both to and from subscribers. Warner Amex's Qube system in Columbus, Ohio, was the country's first interactive cable system. Other Qube systems are now operating in Pittsburgh, Cincinnati, and Houston.

Interconnection A link between the headends of two or more cable systems by microwave, cable relay, or satellite, so that programming can be exchanged, shared, or simultaneously viewed.

Leased Channel Any channel made available by the cable operator for a fee.

Local Origination Programming (LO) Programming developed by an individual cable television system specifically for the community it serves.

Unlike access channels, LO is under the operator's exclusive control. However, LO, public access, and leased access offer many opportunities for the person who wants to appear on cable TV or to produce for it.

MATV (Master Antenna Television System) A system that serves a concentration of television sets—an apartment building, hotel, etc.—utilizing one central antenna to pick up broadcast signals.

MDS (Multipoint Distribution Service) A method of sending TV signals via a form of microwave that beams the signal a short distance in all directions. It is frequently used by pay services to send programming to hotels or to individual homes that are not connected to cable.

Microwave A high-frequency electromagnetic wave used, among other purposes, to relay television signals over long distances via a series of regularly spaced antennas/towers.

MSO (Multiple System Operator) A company that owns and operates more than one cable system.

Narrowcasting Describes a cable system's service to a small community; also used to describe programming that addresses a special-interest group rather than a mass audience. A network such as ESPN has a certain degree of specialization, targeting its programs to an audience interested in sports. In that sense, it can be said that Nickelodeon "narrowcasts." On the other hand, HBO programs for as broad an audience as the commercial networks.

NCTA (National Cable Television Association) The major trade association of the cable television industry, chartered in 1952 with headquarters in Washington, D.C. Thomas Wheeler is the current NCTA president.

New-Build A cable system still in the construction phase, or just recently operational.

Operator The person or company owning a cable system.

Pay Cable Premium services, such as HBO, Showtime, and the Movie Channel, that cost subscribers an additional fee to receive. Movies, sports, and made-for-cable specials and series comprise the pay programming available to subscribers for the extra charge.

Pay Per View (PPV) A form of pay-cable in which a subscriber is charged a separate fee to see an individual program. Examples of pay-per-view shows include the recent Sugar Ray Leonard–Thomas Hearns prize fight, a Rolling Stones concert, and special airings of *Sophisticated Ladies,*

The Pirates of Penzance, and other movies and plays, which the subscriber must pay an additional fee to see.

Penetration   The percentage of total households in a franchise area that are hooked up to cable. The latest statistics indicate that for every ten homes "passed" by cable, five end up subscribing. Thus, the national penetration rate is approximately 50 percent.

Pole Attachments   The cable television hook-ups to telephone and utility poles.

Public Access   The right, often mandated in a franchise agreement, of members of the community to put their own programs on a cable system on a nondiscriminatory basis.

Qube   The Warner Amex two-way cable system first installed in Columbus and since spread to other Warner Amex franchise areas.

Satellite Service   Any channel delivered to cable systems by a communications satellite.

Shop-at-Home   Any program allowing subscribers to view products and/or order them by cable television, including catalogues, shopping shows, and so on.

Spin   An industry term to describe the act of trading one pay service for another. Pay subscribers have begun to channel-shop with abandon; thus, spin is occurring more frequently.

STV   (Subscription TV) Technically, cable television is a "subscription TV" service; STV, however, refers to the delivery of pay programs by over-the-air broadcast methods or MDS. Signals are scrambled and decoded in the subscriber's set by a special receiver.

Subscriber   A person who pay a monthly fee to cable system operators for the capability of receiving cable services.

Superstation   A conventional TV station whose programming is made available to cable systems around the country by satellite. The first was WTCG in Atlanta (now called WTBS), which became a "superstation"— Ted Turner's phrase—in 1976. Less conspicuous superstations include WGN out of Chicago and New York's WOR.

Teletext   A system of storing and displaying printed and graphic material on the home television screen. A one-way transmission of text information.

Tier   A level of program service, usually beginning with a basic tier that comes to the subscriber for his basic fee. Each tier added on to that provides more channels and services on an escalating payment scale.

Transponder   The part on a communications satellite that receives and transmits a signal. There are twenty-four transponders on the newer satellites, allowing each satellite to distribute twenty-four channels of cable programming simultaneously.

Trunk   A cable system's main line, stretching from the headend to the limits of the franchise area.

Two-way Cable   The term used to describe a cable television system that enables signals to pass in both directions, from the headend to the subscriber and back. See also BI-DIRECTIONAL CABLE and INTERACTIVE CABLE.

Underground Installation   A method of installing cable underground, as opposed to aerial suspension on poles.

Uplink   Earth station that sends TV signals to a satellite.

Upstream   The flow of signals in a cable system from subscribers to the headend.

Videotex   An information delivery system defined as a two-way interactive text service that offers, among other subjects, news, sports, home shopping, health and educational information.

Appendix: Cable Directory

CABLE NETWORKS

AETN (American Educational Television Network)
2172 Dupont Dr.   Suite 7   Irvine, CA 92715
714-955-3800
  Programming geared toward continuing education for licensed professionals.

ARTS (Alpha Repertory Television Service)
1211 Sixth Ave.   15th Fl.   New York, NY 10036
212-944-4265
  A joint venture of Hearst/ABC, offering music, dance, and drama, with an emphasis on performance programs.

BET (Black Entertainment Television)
1050 31st St. N.W.   2nd Fl.   Washington, DC 20007
202-337-5260
  Programming oriented toward the black audience—feature films, musical specials, quiz shows, and sports.

BRAVO
100 Crossways Park West   Suite 200   Woodbury, NY 11797
516-364-2222
  Film, dance, and music presentations with a highbrow appeal.

CBN (Christian Broadcasting Network)

CBN Center   1000 Centerville Turnpike   Virginia Beach, VA
23463   804-424-7777

  Religious programming, including soap opera *(Another Life),* college
football, and *The 700 Club.*

CHN (Cable Health Network)

9356 Little Santa Monica Blvd   Beverly Hills, CA 90210
213-550-7241

1211 Sixth Ave.   New York, NY 10036   212-719-7230

  Programming encompasses health, medical, science, and lifestyle topics.

CINEMAX

Time-Life Building   Rockefeller Center   New York, NY 10020
212-484-1100

  A sister service to HBO; mostly movie programming, and a few series,
too.

CNN (Cable News Network)

1050 Techwood Dr.   Atlanta, GA 30318   404-898-8500

  An in-depth news service offering a variety of special-interest programs
and features, in addition to ongoing newscasts.

CNN HEADLINE NEWS

1050 Techwood Dr.   Atlanta, GA 30318   404-898-8500

  A fast-paced headline news service, programmed in half-hour cycles.

C-SPAN (Cable Satellite Public Affairs Network)

400 N. Capitol St. N.W.   Suite 155   Washington, DC 20001
202-737-3220

  Public affairs programming, including coverage of congressional de-
bates, Senate and House committee hearings, and viewer call-in shows
with policymakers in Washington.

DAYTIME

555 Fifth Ave.   New York, NY 10017   212-661-4500

  General-interest programming geared toward women.

THE DISNEY CHANNEL

645 Madison Ave.   New York, NY 10022   212-688-8431

  Family entertainment programming, from Mickey and Pluto.

DON KING SPORTS & ENTERTAINMENT NETWORK
777 Silver Spur Rd.   Suite 230   Rolling Hills Estates, CA 90274
213-377-3007
  Sports, concerts, and special presentations on a pay-per-view basis.

DOW JONES CABLE NEWS
P.O. Box 300   Princeton, NJ 08540   609-452-2000
  Alpha-numeric business news in 15-minute cycles.

EPISCOPAL TELEVISION NETWORK
Box 2060   New York, NY 10163   212-888-0591
  Religious programming.

EROS
2 Lincoln Sq.   New York, NY 10033   212-595-7900
  Adult programming featuring R-rated films.

ESPN (Entertainment & Sports Programming Network)
ESPN Plaza   Bristol, CT 06010   203-584-8477
  Round-the-clock sports events and sports-related programming.

EWTN (Eternal Word Television Network)
5817 Old Leeds Rd.   Birmingham, AL 35210   205-956-9537
  Catholic programming.

GALAVISION
250 Park Ave.   New York, NY 10017   212-953-7550
  Movies, variety, and sports for Spanish-speaking cable subscribers.

HOME BOX OFFICE
1271 Sixth Ave.   New York, NY 10020   212-484-1000
  Cable's first pay service, offering movies, sports, specials, and
series.

HTN PLUS (Home Theater Network)
90 Park Ave.   New York, NY 10016   212-983-5183
  A film and travel service for the entire family.

THE MOVIE CHANNEL
1133 Sixth Ave.   New York, NY 10036   212-944-5352
  24 hours of movies, every day of the week.

MSN (Modern Satellite Network)
45 Rockefeller Plaza   Suite 1460   New York, NY 10111
212-765-3100
  Public service and informational programming, includes *The Home Shopping Show.*

MTV (Music Television)
1133 Sixth Ave.   18th Fl.   New York, NY 10036
212-944-5398
  Clips, movies, and reports featuring contemporary artists and the music they make; in stereo.

THE NASHVILLE NETWORK
2806 Opryland Dr.   Nashville, TN 37214   615-889-6840
  Country music and country humor.

NCN (National Christian Network)
1150 W. King St.   Cocoa, FL 39222   305-632-1510
  Multi-denominational religious programming.

NICKELODEON
1133 Sixth Ave.   24th Fl.   New York, NY 10036
212-944-5521
  Non-violent, pro-social programming for children and adolescents.

NJT (National Jewish Television)
2621 Palisades Ave.   Riverdale, NY 10463   212-549-4160
  Programming for the Jewish community.

OVATION
6464 Sunset Blvd   Suite 880   Hollywood, CA 90028
213-462-7426
  PBS-style programming, available six hours per week on the USA Cable Network. Formerly called the "English Channel."

THE PLAYBOY CHANNEL
8560 Sunset Blvd   Los Angeles, CA 90069   213-659-4080
  Adult films, specials, and series.

PLAYCABLE
1775 Broadway   22nd Fl.   New York, NY 10019
212-708-7900
An all-video-games cable network.

PRIVATE SCREENINGS
330 W. 42nd St.   New York, NY 10036   212-563-2323
R-rated entertainment.

PTL (Praise The Lord)
7224 Park Rd.   Charlotte, NC 28279   704-542-6000
Christian entertainment, news, and specials.

SHOWTIME
1633 Broadway   New York, NY 10019   212-708-1600
10900 Wilshire Blvd   Los Angeles, CA 90024   213-208-2340
A wide variety of programming, including movies, specials, series, and
documentaries.

SIN (National Spanish Television Network)
250 Park Ave.   New York, NY 10017   212-953-7507
Spanish-language series, specials, and news.

SNC (Satellite News Channel)
41 Harbor Plaza Dr.   P.O. Box 10210   Stamford, CT 06904
203-964-8355
A news headline service, continuously updated every 18 minutes.

SPN (Satellite Program Network)
P.O. Box 45684   Tulsa, OK 74145   918-481-0881
Family entertainment with an emphasis on programming for women.

THE SPORTS NETWORK
P.O. Box 10210   41 Harbor Plaza Dr.   Stamford, CT 06904
203-965-6000
Cable's first multi-regional pay sports service.

SPOTLIGHT
2951 28th St.   Suite 2000   Santa Monica, CA 90405
213-450-6488
A mostly movies pay service.

TBN (Trinity Broadcasting Network)
P.O. Box A   Santa Ana, CA 92711   714-832-2950
  Religious programming.

TELEFRANCE USA
1966 Broadway   New York, NY 10023   212-877-8900
  French-language programming with subtitles, available seven days per week over SPN.

USA CABLE NETWORK
208 Harristown Rd.   Glen Rock, NJ 07452   210-445-8550
  Broad-based programming for special-interest groups; divided into four programming blocks.

THE WEATHER CHANNEL
2625 Cumberland Pkwy   Suite 450   Atlanta, GA 30339
404-434-6800
  All weather, all the time.

WGN
5200 S. Harvard   Suite 215   Tulsa, OK 74135
800-331-4806
  Chicago's independent stations, satellite-fed.

WOR
3 Northern Concourse   P.O. Box 4872   Syracuse, NY 13221
315-455-5955
  The nation's station from New York City.

WTBS
1050 Techwood Dr.   Atlanta, GA 30318   404-898-8500
  The original superstation.

PLANNED CABLE SERVICES

ABC/GETTY PAY SPORTS SERVICE
1330 Sixth Ave.   New York, NY 10019   212-887-7297
  An all-sports pay-per-view network.

ALOHA NETWORK
8730 W. Third St.   Los Angeles, CA 90048   213-275-4558
  Programming produced in Hawaii and transmitted to the mainland.

BUENA VISION CHANNEL
912 N. Eastern Ave.   Los Angeles, CA 90063   213-267-1461
  Programming aimed at English-speaking Hispanic households.

THE CHANNEL BLACK
225 Central Park West   Suite 1123   New York, NY 10024
212-875-1468
  Black-oriented theater, music, sports, and drama.

GAMES NETWORK
637 S. Lucerne St.   Los Angeles, CA 90005   213-932-1950
  Video games for the entire family.

KIDVID NETWORK
342 Madison Ave.   New York, NY 10173   212-867-1700
  Daytime programming for children ages 2–11.

LEGAL NEWS NETWORK
1411 Classen Blvd   Oklahoma City, OK 73106   405-528-7019
  Programming for the Perry Mason in all of us.

MAGICABLE
1350 Sixth Ave.   New York, NY 10019   212-247-1700
  Entertainment programming.

PENTHOUSE CABLE NETWORK
909 Third Ave.   20th Fl.   New York, NY 10022
212-593-3301
  Adult programming.

SATELLITE NEWS CHANNEL II
41 Harbor Plaza Dr.   Stamford, CT 06904   203-964-8355
  Feature-oriented news service to supplement SNC.

THE SILENT NETWORK
P.O. Box 1902   Beverly Hills, CA 90213   213-654-6972
  Originally-produced programming for the hearing-impaired viewer, including sign, voice, and captioning.

SPANISH UNIVERSAL NETWORK

2990 Richmond Ave.   Houston, TX 77098   713-522-2013
   Films, news, and entertainment from six Spanish-speaking countries.

SPN/CLASSIC MOVIE CHANNEL

P.O. Box 45684   Tulsa, OK 74145   918-481-0881
   Classic movies

UTV CABLE NETWORK

2208 Route 208   Fair Lawn, NJ 07410   201-794-3660
   Programming that encourages audience involvement; games, home
shopping, and how-to.

TWENTY TOP-RANKED U.S. CABLE SYSTEMS

System (Major Area Served)	Subscribers
1. Cox Cable San Diego (San Diego, CA)	225,000
2. Cablevision Systems (Woodbury, NY)	192,000
3. Warner Amex(Houston, TX)	182,600
4. Manhattan Cable (New York, NY)	174,415
5. Rogers UA Cablevision (San Antonio, TX)	164,745
6. McLean Hunter Cable (West Orange, NJ)	149,990
7. Rogers UA Cablevision (Wayne, NJ)	123,095
8. United Cable (Tulsa, OK)	106,834
9. New York Times Cable (Audubon, NJ)	105,332
10. Prime Cable Corporation (Erie County, NY)	100,287
11. ATC (Rochester, NY)	100,000
12. Gill Cable (San Jose, CA)	98,000
13. ATC (Austin, TX)	97,425
14. Warner Amex (Cincinnati, OH)	95,000
15. Cox Cable of Tidewater (Norfolk, VA)	91,393
16. Caltech Cable (Baltimore, MD)	90,000
17. Group W of Seattle (Seattle, WA)	87,618
18. Viacom of Long Island (Central Islip, NY)	87,121
19. Orange/Seminole Cablevision (Orlando, FL)	86,700
20. Memphis CATV (Memphis, TN)	85,025

FIFTY LARGEST CABLE SYSTEM OPERATORS

1. American TV & Communications Corp. (ATC)
   160 Inverness Drive West   Englewood, CO 80112
   303-773-3411

2. Tele-Communications, Inc. (TCI)
   5455 S. Valentia Way   Englewood, CO 80111
   303-771-8200

3. Group W Cable
   888 Seventh Ave.   New York, NY 10016   212-247-8700

4. Cox Cable Communications Inc.
   219 Perimeter Center Pkwy   Atlanta, GA 30346
   404-393-0480

5. Storer Cable Communications
   1177 Kane Concourse   Miami Beach, FL 33154
   305-866-0211

6. Warner Amex Cable Communications Inc.
   75 Rockefeller Plaza   New York, NY 10019
   212-484-8000

7. Times-Mirror Cablevision
   2381 Morse Ave.   Box 19398   Irvine, CA 92713
   714-549-2173

8. Newchannels Corporation
   3 Northern Concourse   P.O. Box 4872   Syracuse, NY 13221
   315-455-5971

9. Viacom Communications
   1211 Sixth Ave.   New York, NY 10036   212-575-5175

10. Rogers/UA Cablesystems
    315 Post Road West   Westport, CT 06880   203-227-9581

11. Continental Cablevision Inc.
    Lewis Wharf, Pilot House   Boston, MA 02110
    617-742-9500

12. United Cable TV Corporation
Terminal Annex, Box 5840   Denver, CO 80217
303-779-5999

13. Sammons Communications
500 S. Ervay   Box 225728   Dallas, TX 75002
214-742-9828

14. Telecable Corporation
740 Duke St.   Box 720   Norfolk, VA 23510
804-446-2565

15. Capital Cities Communication
7120 East Orchard Rd.   Englewood, CO 80111
303-770-7500

16. General Electric Cablevision Inc.
1400 Balltown Rd.   Schenectady, NY 12309
518-385-1368

17. Daniels & Associates Inc.
2930 E. Third Ave.   Box 6008   Denver, CO 80206
303-321-7550

18. Comcast Corporation
1 Belmont Ave.   Suite 227   Bala Cynwyd, PA 19004
215-667-4200

19. Heritage Communications Inc.
2195 Ingersoll Ave.   Des Moines, IA 50312
515-245-7585

20. Cablevision Systems Development Company
1 Media Crossways Dr.   Woodbury, NY 11797
516-364-8450

21. Liberty Communications Inc.
2225 Coburg Rd.   Eugene, OR 94701   503-485-5611

22. Wometco Communications Inc.
316 N. Miami Ave.   Miami, FL 33128   305-579-1200

23. Service Electric Cable TV Inc.
1043 Hamilton St.   Allentown, PA 18101   215-827-7750

24. Tele-Media Corporation
P.O. Box 39   Bellefont, PA 16823   814-237-1512

25. Texas Community Antennas Group
P.O. Box 6840   Tyler, TX 75711   214-595-3701

26. Multimedia Cablevision Inc.
140 W. 9th St.   Cincinnati, OH 45202   513-352-5000

27. Century Communications Corporation
51 Locust Ave.   New Canaan, CT 06840   203-966-8746

28. Jones Intercable Inc.
5275 DTC Pkwy   Englewood, CO 80111   303-740-9700

29. Communications Services Inc.
100 Rimrock   Box 829   Junction City, KS 66441
913-762-2570

30. Colony Communications Inc.
P.O. Box 969   Providence, RI 02901   401-277-7444

31. Western Communications Inc.
P.O. Box 4610   Walnut Creek, CA 94596   415-935-3055

32. Harron Communications Corporation
2063 Suburban Station Bldg   Philadelphia, PA 19103
215-569-2935

33. MacLean Hunter
27 Faskin Dr.   Rexdale, Ontario, Canada M9W1K7
416-675-5930

34. Prime Cable Corporation
1515 City National Bank Bldg   Austin, TX 78701
512-476-7888

35. Centel Communications Corporation
993 Oak St.   Aurora, IL 60506   312-897-2288

36. Harris Cable Corporation
10889 Wilshire Blvd   Suite 1240   Los Angeles, CA 90024
213-208-6118

37. Tribune Cable Corporation
Colonial Offices   Whitney Rd   Mahwah, NJ 07430
201-891-7988

38. Midwest Video Corporation
860 Tower Bldg   Little Rock, AR 72201   501-375-8885

39. Cablevision Industries Inc.
P.O. Box 311   Weirk Ave.   Liberty, NY 12754
914-292-7550

40. Rollins Inc.
2170 Piedmont Rd. N.E.   Atlanta, GA 30324
404-873-2355

41. McCaw Communications Company, Inc.
2000 110th Ave.   P.O. Box 3867   Bellevue, WA 98009
206-453-1115

42. Midcontinent Cable, Inc.
24 First Ave. N.E.   P.O. Box 910   Aberdeen, SD 57401
605-229-1775

43. New York Times Cable
120 W. Merchant St.   Audubon, NJ 08106   609-547-4400

44. Communications Systems
700 West Airport Fwy   Suite 790   Irving, TX 75061
214-438-9450

45. Sutton Capitol Group
231 E. 35th St.   New York, NY 10016   212-686-8022

46. Gill Cable, Inc.
1302 N. 4th St.   San Jose, CA 95112   408-998-7333

47. Armstrong Utilities
1 Buckeye Place   Butler, PA 16001   412-283-0925

48. Televents, Inc.
2855 Mitchell Dr.   Suite 250   Walnut Creek, CA 94598
415-935-8010

49. McDonald Group
  P.O. Box 43606   1 Office Park Circle   Suite 300
  Birmingham, AL 35243   205-967-8732

50. Adams-Russell
  1380 Main St.   Waltham, MA 02154   617-894-8540

CABLE TRAINING SCHOOLS

ARIZONA

Maricopa County Skills Center
4118 East Wood St.   Phoenix, AZ 85040   602-243-4141
  Type of program: Cable television installation and technology
  Length of program: 12–26 weeks

CALIFORNIA

Cox Cable Communications
1331 North Cuyamuaca   Suite P   El Cajon, CA 92020
714-562-0742
  Type of program: Cable television system plant maintenance; cable television sweep plant maintenance
  Length of program: Plant maintenance—2 weeks; Sweep plant maintenance—1 week

East Bay Skills Center
1100 67th St.   Oakland, CA 94608   415-658-7256
  Type of program: Cable installer/technician
  Length of program: 7 months

San Diego Community College
12th & B Sts.   San Diego, CA 92101   714-562-1150
  Type of program: Cable television technology
  Length of program: 2 semesters

COLORADO

ATC National Training Center
2100 S. Hudson   Denver, CO 80222   303-773-3411
  Type of program: Installer, technician I and II
  Length of program: Approximately 1 year

National Cable Television Institute
P.O. Box 27277   Denver, CO 80227   303-697-4967
  Type of program: Correspondence courses in cable television technical
training
  Length of program: 6 months–3 years

RETS Electronic Schools
7346 S. Alton Way   Suite A   Englewood, CO 80112
303-741-5873
  Type of program: Installer, construction, basic–chief technician
  Length of program: 4 weeks–2 months

CONNECTICUT

Middlesex Community College
100 Training Hill Rd.   Middleton, CT 06457   203-344-3052
  Type of program: Associate degree in cable telecommunications
  Length of program: 2 years

DISTRICT OF COLUMBIA

McGraw-Hill Continuing Education Center
3939 Wisconsin Ave. N.W.   Washington, DC 20016
202-244-1600
  Type of program: Correspondence, cable technician I and II
  Length of program: 18 months

ILLINOIS

Omega School of Communications
548 N. Lake Shore Dr.   Chicago, IL 60611   312-321-9400
  Type of program: FCC license program
  Length of program: 10 weeks

MINNESOTA

Dakota County Area Vocational Technical Institute
Student Services Office   145th St. E. & Akron Rd.   P.O. Drawer K
Rosemont, MN 55068   612-423-2281
Type of program: Cable television technician
Length of program: 22 months

Wadena Area Vocational Technical Institute
P.O. Box 267   405 S.W. Colfax Ave.   Wadena, MN 56482
            218-631-3530
Type of program: Cable television technology
Length of program: 22 months

MISSOURI

Chillicothe Area Vocational Technical School
1200 Fair St.   Chillicothe, MO 64601   816-646-3414
Type of program: Electronics technician
Length of program: 2 years (1080 hours)

National Cable Training Center
4150 Old Mill Pkwy   St. Charles, MO 63301   314-441-7490
Type of program: Cable TV installer/technician and linemen
Length of program: 7–21 weeks

NEW JERSEY

Essex County College
303 University Ave.   Newark, NJ 07102   201-877-3274
Type of program: Associate degree program in broadcast technology
and operations
Length of program: 2 years

Mercer County Community College
1200 Old Trenton Rd.   Trenton, NJ 08690
609-586-4800 (ext. 463)
Type of program: Associate degree program in telecommunications
technology
Length of program: 2 years

NEW YORK

BOCES Electronics
Boces Nike Center   Shawnee Rd.   Sanborn, NY 14132
716-731-9620
Type of program: Electronics cable technician
Length of program: 750 hours

OHIO

Cleveland Institute of Electronics
1776 E. 17th St.   Cleveland, OH 44114   216-781-9400
Type of program: Correspondence course in electronics
Length of program: Varies

Garfield Skills Center
2013 W. Third St.   Dayton, OH 45417   513-268-6702
Type of program: Cable television technology
Length of program: 1 year

University College/University of Cincinnati
M.L. 47   Cincinnati, OH 45221   513-475-3551
Type of program: Associate degree
Length of program: 2 years

OKLAHOMA

South Oklahoma City Junior College
7777 S. May Ave.   Oklahoma City, OK 73159   405-682-1611
Type of program: Associate degree program in electronics engineering
with a cable television concentration
Length of program: 2 years

PENNSYLVANIA

International Correspondence Schools
Oak Street   Scranton, PA 18515   717-342-7701
Type of program: Independent study leading to an associate degree in
electronics technology
Length of program: Varies

TEXAS

Texas Engineering Extension Service/Electronics Training Division
Texas A & M University System
F.E. Drawer K   College Station, TX 77843
713-779-3880 (ext. 244)
  Type of program: Cable TV installer, technician, chief technician
  Length of program: 5 days (40 hours) for each program

WISCONSIN

Western Wisconsin Technical Institute
Sixth & Vine Sts.   La Crosse, WI 54601   608-785-9178
  Type of program: Installer/maintenance; installer/technician
  Length of program: 7 weeks; 20 weeks

Wisconsin Indianhead Technical Institute
Rice Lake Campus   1900 College Dr.   Rice Lake, WI 54868
715-234-7082
  Type of program: One year vocational diploma
  Length of program: 9 months (August–May)

PROFESSIONAL ASSOCIATIONS

There are several associations in the cable television industry which allow
nonprofessionals to enter as a subscribing or an associate member. Members receive many benefits (newsletters, seminars, workshops, career guidance, etc.) and the chance to network their way into a job. Contact the
professional associations listed below and ask for their literature.

CABLE TELEVISION ADMINISTRATION AND MARKETING SOCIETY (CTAM)
219 Perimeter Center Pkwy   Suite 480   Atlanta, GA 30346
404-399-5574
  The association holds educational seminars at cable conventions across
the country, and offers its own annual convention open to members and
non-members alike. Members receive a quarterly newsletter, which includes a listing of cable events region by region. Membership is currently
$50 per year.

CABLE TELEVISION ADVERTISING BUREAU
767 Third Ave.   New York, NY 10017   212-751-7770

Established in 1981, this trade association counts among its members cable systems, cable networks, sales representatives, and consultants—all interested in the growing importance of cable as an advertising medium. Since there is a need for energetic and talented people to sell local spots on the system level and national spots on the network level, CTAB is a good place to make important contacts and to hear more about opportunities. The association holds an annual conference, usually in the early spring, and currently offers free of charge an information packet, which includes articles on how to be a cable salesperson.

CABLE TELEVISION INFORMATION CENTER (CTIC)
1800 N. Kent St.   Suite 1007   Arlington, VA 22209
703-528-6846

Founded in 1972, this association provides consulting services to the cable industry, with the emphasis on local policies, regulations, and franchising. The association also provides help for people who desire information on access programming. An annual membership, which includes a subscription to the monthly newsletter, is $165.

COMMUNITY ANTENNA TELEVISION ASSOCIATION (CATA)
3977 Chain Bridge Rd.   Fairfax, VA 22030   703-691-8875

This trade association publishes a monthly newsletter and a monthly magazine, and organizes a seminar which is held once a year, usually in August. The association is also involved with regional seminars, as many as twelve a year, which delve into the technical side of cable TV. A one-year individual membership costs $50.

FOUNDATION FOR COMMUNITY SERVICE CABLE TV
5616 Geary Blvd   Suite 212   San Francisco, CA 94121
415-387-0200

For access enthusiasts in the state of California, the state-mandated foundation serves as a clearinghouse of information, and helps promote and encourage the use of access channels by individuals, municipal groups, educators, and others. Founded in 1979, the foundation will provide an information packet and a quarterly newsletter to interested callers, and help identify support clubs and cable organizations in the caller's area. All free of charge.

NATIONAL CABLE TELEVISION ASSOCIATION (NCTA)
1724 Massachusetts Ave., N.W.   Washington, DC 20036
202-775-3550

The major trade association in the cable industry, NCTA offers an individual patron membership at $200 per year. Each member is entitled to a subscription to the monthly newsletter, discounts on NCTA publications, and the use of the research departments of the association. Perhaps most important, a membership gains admission to the various NCTA-sponsored conventions and seminars held around the country. It is at the conventions and seminars that valuable contacts can be made.

NATIONAL FEDERATION OF LOCAL CABLE PROGRAMMERS (NFLCP)
906 Pennsylvania Ave. S.E.   Washington, DC 20003
202-544-7272

Established in 1976, this association offers an annual individual membership for $25. As a member of NFLCP, you will receive a monthly magazine called *Community TV Review*—which covers the access scene around the country and job opportunities on the local level—and a monthly newsletter, which lists the numerous regional and local conferences the association sponsors each year. If you're interested in access, as a would-be star, producer, production person, or regulator, the NFLCP is a good place to start.

WOMEN IN CABLE
2033 M St. N.W.   Suite 703   Washington, DC 20036
202-296-4218

A professional society, with membership open to women and men employed in cable television or closely allied fields. The organization has over 2,000 members—95 percent of whom are women—and currently sponsors 25 local chapters across the country. In addition to the local chapter meetings and its annual national convention—where networking can be an art form—Women In Cable publishes a monthly newsletter for the cable-minded, and runs frequent career seminars.

CABLE-RELATED NEWSPAPERS, PERIODICALS, AND DIRECTORIES

If not at your local newsstand, bookstore, or library, contact publishers for availability and cost.

ADVERTISING AGE
740 Rush St.   Chicago, IL 60611
Columns and stories about the cable industry, in addition to a help-wanted section. Published weekly.

AMERICAN FILM: MAGAZINE OF THE FILM AND TELEVISION ARTS
American Film Institute   John F. Kennedy Center for the Performing Arts   Washington, DC 20566
News stories, think pieces, and features that occasionally touch on the cable revolution. Published monthly.

BACKSTAGE
Back Stage Publications   165 W. 46th St.   New York, NY 10023
A newspaper edited for the entire communications entertainment industry. Includes career information for behind-the-scenes personnel. Published weekly.

BROADCAST COMMUNICATIONS
4121 W. 83rd St.   Suite 132   Prairie Village, KS 66208
A magazine for the hard-core broadcaster and cablephile. Published monthly.

BROADCAST ENGINEERING
P.O. Box 12901   Overland Park, KS 66212
For the experienced and well-versed engineer. Published monthly.

BROADCASTING
1735 DeSales St. N.W.   Washington, DC 20036
Contains stories about cable TV; help-wanted section at the back. Required reading. Published weekly.

BROADCASTING'S CABLECASTING SOURCEBOOK.
1735 De Sales St. N. W.   Washington, DC 20036
A comprehensive annual directory for the cable-minded, which lists cable systems, suppliers, and other pertinent names and numbers.

CABLEAGE
1270 Sixth Ave.   New York, NY 10020
A magazine that keeps up-to-date on the technology and people that drive the industry. Published weekly.

CABLE COMMUNICATIONS
4 Smethana Dr.   Kitchener, Ontario, Canada N2B3B8

CABLE MARKETING
352 Park Ave. S.   New York, NY 10016
  A magazine that keeps close tabs on the selling, and the salespeople, of cable TV. Published monthly.

CABLE NEWS
7315 Wisconsin Ave.   Suite 1200 N   Bethesda, MD 20814

CABLE TELEVISION BUSINESS
6430 S. Yosemite St.   Englewood, CO 80111
  One of the most thorough magazines available; contains a very helpful help-wanted section. Required reading. Published monthly.

CABLEVISION
2500 Curtis St.   Suite 200   Denver, CO 80205
  A thoroughly well-done magazine that stays on top of the trends and tribulations of the cable world, and includes a help-wanted section, events calendar, and cable stats list in the back. Required reading for people who want to learn. Published weekly.

CATJOURNAL
4209 N.W. 23rd   Suite 109   Oklahoma City, OK 73101
  Community antenna television journal which covers the cable industry. Published quarterly.

CHANNELS
1515 Broadway   New York, NY 10036
  Contains think pieces and profiles on television and the people who have a say. Published quarterly.

COMMUNICATIONS DAILY
1836 Jefferson Pl. N.W.   Washington, DC 20036
  Local, national, and international coverage. Published daily.

COMMUNICATIONS NEWS
124 S. First St.   Geneva, IL 60134

DAILY VARIETY
1400 N. Cahuenga Blvd　　　Hollywood, CA 90028

A newspaper that endeavors to provide complete objective news as well as news analysis of the entertainment business, including cable TV. Lists job availabilities every day. Published daily except Saturdays, Sundays, and holidays.

FILMMAKERS MONTHLY
P.O. Box 115　　　Ward Hill, MA 08130

A magazine for those involved in feature film and video independent production. "Bulletin Board" section announces jobs wanted and available. Published monthly.

THE HOLLYWOOD REPORTER
6715 Sunset Blvd　　　Hollywood, CA 90028

Provides national and international coverage of the entertainment industry, with a couple of fast-paced columns included. Also a help-wanted column every day. A must for those living in New York or Los Angeles, and for those planning to arrive soon. Published daily except Saturdays, Sundays, and holidays.

HOME VIDEO
474 Park Ave. S.　　　New York, NY 10016

A magazine devoted to the evolving home entertainment industries, including video discs, cassettes, and the new technologies. Published monthly.

HOME VIDEO & CABLE REPORT
701 Westchester Ave.　　　White Plains, NY 10604

PAUL KAGAN REPORTS
26286 Carmel Rancho Lane　　　Carmel, CA 93923

A series of newsletters covering cable franchising, pay-TV programming, cable finance, cable marketing, and other areas of interest to the cable professional. Published biweekly and monthly.

MARKETING COMMUNICATIONS
475 Park Ave. S.　　　New York, NY 10016

MILLIMETER
12 E. 46th St.　　　New York, NY 10017

A magazine that covers theatrical motion pictures, broadcast TV, TV commercials, and cable, with an emphasis on the people behind the scenes. Published monthly.

MULTI-CHANNEL NEWS
300 S. Jackson St.   Denver, CO 80219

A newspaper that blankets the cable industry and provides an important help-wanted section in each issue. Required reading. Published weekly.

SAT GUIDE
P.O. Box 1048   Hayley, ID 83333

SAT stands for "satellite" and, appropriately enough, covers the goings-on in the cable satellite business. Published monthly.

SHOW BUSINESS
134 W. 44th St.   New York, NY 10036

A newspaper edited essentially for the out-of-work stage actor and actress, it includes lists of agents, names of shows in production, and times of casting calls. Occasionally gives a nod to cable TV. Published weekly.

TELEVISION DIGEST FACTBOOK
1836 Jefferson Pl. N.W.   Washington, DC 20036

An annual sourcebook with a plethora of cable data.

VARIETY
154 W. 46th St.   New York, NY 10036

The weekly newspaper, edited for those in the entertainment profession, including motion pictures, live theater, home video, broadcast television, and cable TV. A section is devoted each week to the happenings in the cable industry. Required reading.

VIDEOGRAPHY
474 Park Ave. S.   New York, NY 10016

For people who take the production side of the business seriously. Published monthly.

VIEW
150 E. 58th St.   New York, NY 10022

A magazine that explores the world of cable programming in a lively format which includes feature stories, columns, and news items. Required reading for people who are getting an education in cable TV. Published monthly.

INDEX

(*Italic* numbers refer to major discussions)